Technikzukünfte, Wissenschaft und Gesellschaft / Futures of Technology, Science and Society

Reihe herausgegeben von
A. Grunwald, Karlsruhe, Deutschland
R. Heil, Karlsruhe, Deutschland
C. Coenen, Karlsruhe, Deutschland

Diese interdisziplinäre Buchreihe ist Technikzukünften in ihren wissenschaftlichen und gesellschaftlichen Kontexten gewidmet. Der Plural „Zukünfte" ist dabei Programm. Denn erstens wird ein breites Spektrum wissenschaftlich-technischer Entwicklungen beleuchtet, und zweitens sind Debatten zu Technowissenschaften wie u.a. den Bio-, Informations-, Nano- und Neurotechnologien oder der Robotik durch eine Vielzahl von Perspektiven und Interessen bestimmt. Diese Zukünfte beeinflussen einerseits den Verlauf des Fortschritts, seine Ergebnisse und Folgen, z.B. durch Ausgestaltung der wissenschaftlichen Agenda. Andererseits sind wissenschaftlich-technische Neuerungen Anlass, neue Zukünfte mit anderen gesellschaftlichen Implikationen auszudenken. Diese Wechselseitigkeit reflektierend, befasst sich die Reihe vorrangig mit der sozialen und kulturellen Prägung von Naturwissenschaft und Technik, der verantwortlichen Gestaltung ihrer Ergebnisse in der Gesellschaft sowie mit den Auswirkungen auf unsere Bilder vom Menschen.

This interdisciplinary series of books is devoted to technology futures in their scientific and societal contexts. The use of the plural "futures" is by no means accidental: firstly, light is to be shed on a broad spectrum of developments in science and technology; secondly, debates on technoscientific fields such as biotechnology, information technology, nanotechnology, neurotechnology and robotics are influenced by a multitude of viewpoints and interests. On the one hand, these futures have an impact on the way advances are made, as well as on their results and consequences, for example by shaping the scientific agenda. On the other hand, scientific and technological innovations offer an opportunity to conceive of new futures with different implications for society. Reflecting this reciprocity, the series concentrates primarily on the way in which science and technology are influenced social and culturally, on how their results can be shaped in a responsible manner in society, and on the way they affect our images of humankind.

Weitere Bände in der Reihe http://www.springer.com/series/13596

Martin Sand

Futures, Visions, and Responsibility

An Ethics of Innovation

 Springer VS

Martin Sand ⓘD
Karlsruhe, Germany

Dissertation Karlsruhe Institute of Technology, Germany, 2018

ISSN 2524-3764   ISSN 2524-3772   (electronic)
Technikzukünfte, Wissenschaft und Gesellschaft / Futures of Technology, Science and Society
ISBN 978-3-658-22683-1   ISBN 978-3-658-22684-8   (eBook)
https://doi.org/10.1007/978-3-658-22684-8

Library of Congress Control Number: 2018948630

Springer VS

Printed on acid-free paper

This Springer VS imprint is published by the registered company Springer Fachmedien Wiesbaden GmbH part of Springer Nature
The registered company address is: Abraham-Lincoln-Str. 46, 65189 Wiesbaden, Germany

To K.J.

Acknowledgments

The Department of Social Sciences and Humanities at Karlsruhe Institute of Technology (KIT) has accepted this book in a slightly revised version as dissertation in December 2017. The dissertation was defended on January 22nd 2018.

Science is a team effort and this work is no exception in this regard. Without the continuous support of my supervisors, colleagues, friends and family, this book would have a fundamentally different (and most certainly inferior) character. Prof. Armin Grunwald, my supervisor at the Institute of Technology Assessment (ITAS) in Karlsruhe, has been a mentor in a truly Homerian sense: He accompanied this project with affirmation, continuous support and constructive criticism. I am extremely grateful to him for having been granted the liberty to explore this field from rather unusual directions. Armin Grunwald's sustained trust in this project, which illustrates a virtue following the theory of this dissertation, encouraged me through difficult times. Prof. Ibo van de Poel from Delft University of Technology joined this project when it was already in an advanced stage and yet he was enthusiastic and supportive from the very beginning. Without hesitation, Ibo embraced his role as a second supervisor, made helpful comments on the manuscript and travelled to Karlsruhe for the defence. I am looking forward to our future collaborations and I hope that we keep on challenging each other's views about collective and individual responsibility.

Aside from the financial support to present this work at a number of conferences and to acquire additional academic and language skills, the ITAS has provided a great research infrastructure, an exciting interdisciplinary environment and an open and familial atmosphere ever since I began working there as a student researcher in 2011. There is no adequate English translation of the German word "Heimat". If such term can be applied in this context, it would be the most appropriate description of my relation to this institution and the people who compose it. This project has benefitted greatly from discussions with numerous colleagues. I am particularly indebted to Julia Hahn, Daniel Frank, René König, Reinhard Heil, Christoph Schneider, Christopher Coenen, Arianna Ferrari, Klaus Wiegerling, Michael Poznic, Karsten Bolz, Stefan Böschen and all the members of the project group "Visions as socio-epistemic practices". I already miss the PhDs at ITAS, too many to mention all of them by name, who arranged hilarious lunch sessions at the Moltke-Mensa and other enjoyable social events. The research assistants at ITAS, particularly Elke Träutlein, Muazez Genc and Claudia Lange, contributed to creating an outstanding working environment to undertake such project without friction. Because of their assistance, my cluelessness in administrative regards never became public. Lorie Schiesl and Megan Tennant have done a superb job in helping me to improve my English. If this book is still unreadable, it is entirely my fault.

Prof. Wolfgang Neuser from the Technical University Kaiserslautern gave me the opportunity to engage with students from various backgrounds to discuss issues in business ethics from 2015-2017. The Corporate State University Baden-Württemberg also gave me the opportunity to discuss my thoughts on ethics, ageing and technology and to receive critical feedback from eager students. Both seminars have been enormously valuable experiences. Being spurred to learn things for explaining them to non-philosophers, raised my awareness for many hidden facets of moral philosophy.

I am greatly indebted to the Karlsruhe House of Young Scientists (KHYS), which rewarded me with a Networking Grant to discuss my ideas with the research group of Prof. Harro van Lente at the Department of Technology and Society Studies at Maastricht University in May 2015. Furthermore, the KHYS generously supported a five-month research visit at the Department of Values, Technology and Innovation at Delft University of Technology in the beginning of 2018, which allowed me to explore new perspectives on responsibility and further deepen the ideas outlined in this book. These international experiences have been invaluable contributions to my personal and philosophical development.

Both the Department of Values, Technology and Innovation in Delft and the Department of Technology and Society Studies in Maastricht have provided an incredibly hospitable and inspiring atmosphere, which contributed to the success of this project. The German Academic Exchange Service (DAAD) generously funded a trip to the S.NET Conference in Arizona in 2017, where I presented some ideas from this book. In a slightly revised version, Chapter seven of this book received the Best Paper Award at the Philosophy of Management Conference in Oxford in 2016. Springer generously contributed to this prize with a book voucher.

I am extremely grateful for having received this honour. After submitting this thesis, I was fortunate to receiving a Fellowship as Visiting Researcher to discuss my ideas with the members of the Graduate School "Innovation Society: Today" at Technical University Berlin. This provided me with valuable insights into the sociology of technology and sociological theories of agency.

Since I exist, my parents, Rosi and Herbert Sand, supported me without doubt or hesitation. Words cannot express how much I owe them. Without knowing or questioning, they have always embraced what my supervisor calls a forward-looking responsibility. In less technical terminology, I simply hope that they are as proud of me, as I am of them. There are numerous friends, whose patience has been tested while this book was in the making. I am extremely fortunate to have them and tremendously grateful that they always return my calls when I need guidance or consolidation. Their persistence helped me to master difficult times, to finish this project successfully and to keep on pursuing my peculiar athletic ambitions. In no particular order, these people include Lisa O., Roman, Simon, Hendrik, Manu, Christian, Lisa A., Colette, Johanna, Claudia, Jonas, Niklas, Noah, Giuli, Matthias, Do, Wouter, Lisanne, Gustavo, Rolanda, Paul, Linda and all others that feel addressed.

This book is dedicated to the person, who had the greatest influence on me in the past five years. Karin is everywhere between the lines of this book. Since I know her, she has been an endless source of inspiration, energy and intellectual fervour. Without her, my work and life would be substantially depleted. I am looking forward to our upcoming adventures and I cannot wait to see how she challenges me next.

Contents

List of Tables and Figures

Abstract

The present enquiry explores problems of responsibility at the early, visionary stages of technological development. After introducing the research question in chapter one, chapter two discusses the increasingly dominant concept of innovation. I understand innovations as novelties that can occur across disciplines, including technological, social, political, and natural realms. I will argue that such qualitative changes are ambivalent. As agency is crucial to bringing about novelty, using evolutionary metaphors as *explanans* for technological development is problematic. The central role of agency in this process constitutes a humanist ethics of innovation. Chapter three continues with an outline of how narratives about the future are currently used to facilitate technological change, particularly with regard to the example of "visioneering," to foster networks, and to raise public awareness for innovations. In this example, necessary conditions for being responsible (intentionality, alternatives, and accountability) are introduced, and I explain why reducing the responsibility of visioneers to their potential effectiveness is insufficient for holding them responsible. In chapters four and five, I discuss the problems of genuine freedom and determinism. A person's ability to act for reasons is a phenomenon *sui generis* that is inexplicable in causal terms. I argue that because of the existence of agency, the theory of universal determinism is implausible. Furthermore, I argue that most forms of compatibilism that rely on weaker notions of agency presuppose a libertarian understanding of agency. By discussing the problem of moral luck and effect compatibilism, these Chapters further advance the concept of responsibility by distinguishing "being responsible" from "being held responsible." Praising and blaming, which are forms of holding agents responsible, are actions that are not always adequate responses when the basic conditions for being responsible are met. Strawson's theory of moral sentiments is discussed as a compatibilist theory. Chapter six considers different theories of collective and corporative responsibility, as visioneers and innovators normally act in relatively structured collective settings. I advance a view that understands collective and corporative responsibility as pragmatic foreshortening. Chapter seven develops a constructive theory of innovators' and visioneers' responsibilities. These societal agents pioneering technological change are responsible for their characters, their approaches to dealing with complexity, and for determining how creative and eager they are to pursue new technological pathways. I argue that such virtues are important for change and that they can be trained like handling a tool. The eligibility of this virtue ethical framework is illustrated through the discussion of two captivating examples–Steve Jobs and Victor Frankenstein.

1. Introduction

*What is significant, what gives any human exist-
ence its meaning, is the possibility that thus arises
of creative power. But it is no more than a possibility, realized here and there, more or less, and fully
realized only in exceptional persons. (Taylor 1987,
685)*

1.1 Research Question

There is a growing interest in the interrelation between responsibility and innovation from an interdisciplinary perspective (Blok and Lemmens 2015; Bogner et al.
2015). New and emerging technosciences such as big data and synthetic biology
challenge existing paradigms of assessing technological systems as being applied,
for instance, in Technology Assessment (TA). Over the past couple of years, these
technosciences motivated the development of a more advanced understanding of
technological change (Nordmann 2011; Grunwald 2015). In order to cope with
these emerging challenges "new paradigms for science, technology and innovation
policy," such as the idea of responsible research and innovation (RRI), and new
forms of assessing the visions of emerging technologies such as hermeneutic TA
have been developed (Grunwald 2013, 2014a; von Schomberg 2015). Following
up on this research, I will discuss the interrelation between responsibility and innovation from a philosophical perspective and the problems of responsibility when
new technologies are *promoted* and *established*. Innovation processes are the
phases in which new technologies are established or qualitative changes in existing
technological systems are brought about. Narratives that express a common technological future and that can persuade others to strive for technological change
often initiate such innovation processes. More generally, such narratives function
as *prisms* to reflect on the feasibility and desirability of novel technologies. Visions
and narratives about the future are understood to be performative in this sense

© Springer Fachmedien Wiesbaden GmbH, part of Springer Nature 2018
M. Sand, *Futures, Visions, and Responsibility*, Technikzukünfte, Wissenschaft
und Gesellschaft / Futures of Technology, Science and Society,
https://doi.org/10.1007/978-3-658-22684-8_1

(Michael 2000, 33; Grunwald 2017). This research aims to contribute to answering the following research question: Do relevant agents in innovation processes, including innovators and people distributing visions and narratives of technological futures, carry a special form of responsibility?

When speaking about responsibility, it should be mentioned in advance that this research is mainly concerned with *moral* responsibility.[1] There are clear overlaps in judgments about legal and moral responsibility in jurisdictions of European and Northern American countries. People who are incapable of making reasonable judgments (children) or controlling their behavior (addicts) are clearly not morally or legally accountable, which is agreed upon by most jurisdictions and ethicists alike (Keil 2011, 157; Bieri 2013, 342). A lack of foreknowledge and bad intentions are also mitigating factors when determining the scope of legal sanctions. This is also taken into account in debates about moral responsibility. Still, there are vast differences between moral and legal responsibility; we can say that there is a greater *widths* to moral responsibility. Moral responsibility transcends the boundaries of positive laws, which are usually the borders of national states. It would be wrong to torture innocent people for no reason, even if there were no complainants, national states, or judges around witnessing the act (say on a foreign planet or in a distant future). In this sense, moral responsibility appears to be associated with the idea of *universal validity* exceeding the temporally and locally restricted "areas" in which positive laws prevail (Nagel 1986, 158; Bieri 2016, 76). This aspect also reminds us of the apparent discrepancies between what is moral and what has been considered legal in times of despotism and tyranny. What has been considered as lawful in Nazi Germany, for example, has never been and will never be morally acceptable. Positive laws can be instantiated with brute force or violence without ever resembling what is morally right and good. Such political systems are not legitimate, but they nevertheless determine the scope of legal responsibility (Nagel 1991a). In contrast, it would in itself be immoral to establish a "moral system" in an unwarranted, violent way. The standard according to which

---

1   I will sometimes refer to the distinction between legal and moral responsibility, for instance, in the context of collective responsibility. There, the existence of corporates' legal liability provides another reason to not forcefully search for grounds to attribute moral responsibility to such bodies (see chapter six).

theories of moral responsibility are being evaluated is not legitimacy but reasonableness. Therefore, a theory of moral responsibility must be supported by good arguments (Nagel 1986, 139; Williams 1997, 18; Nida-Rümelin 2011, 12). As Peter Singer writes in his *Practical Ethics*: "The notion of living according to ethical standards is tied up with the notion of defending the way one is living, of **giving a reason** for it, of **justifying** it [own emphasis]." (Singer 2008, 10) Moral responsibility also has in some respects a greater *depth* than legal responsibility. There are no laws that forbid cowardice or arrogance and yet such traits are morally blameworthy. While laws also determine aspects of everyday life, for example the size of trees one is allowed to grow in a populated neighborhood, they are sometimes without clear moral implications, and the range of judgments in theories of moral responsibility can be much more nuanced and subtle to possibly being employed in contexts of jurisprudence. In order to adhere to this standard for goodness, and to make the viewpoints and the criticism outlined as persuasive and compelling as possible, I will provide *arguments* to defend my theory of moral responsibility in the following chapters. Those arguments will often be exemplified with fictive or real life cases to further strengthen my conclusions. Both methods typically belong to the armamentarium of philosophy.

In order to answer my research question, the concept "innovation" and the problem of agency that innovations' complexity might entail will be discussed in the second chapter. In chapter three I will introduce "visioneering" as a form of agency that affects the innovation processes and discuss arguments which suggest that this type of agent carries a special form of responsibility (Coenen 2011; Cabrera Trujillo, Laura Yenisa 2014; Ferrari and Marin 2014). This discussion will uncover some preconditions for being responsible (intentionality and the existence of alternatives). Furthermore, I will explore the problem of accountability resting on the complexity and opaqueness of innovation journeys (Rip 2012). Chapter four extends the analysis of responsibility and discusses the problem of free will and agency in more detail. Positive accounts of responsibility rely on the notion of people's general ability to perform certain acts instead of other *alternatives*, and the difficulty of giving a more precise account of this notion facilitates a discussion about free will. In this chapter, I will distinguish the idea of being *held responsible* from *being responsible* and offer some examples for these notions and how they can diverge. Chapter five analyses the problem of moral luck which was first discussed with slightly different foci by Bernard Williams and Thomas Nagel (Williams 1981b; Nagel 1991b). Given the unpredictability and opaqueness of innovation journeys, the problem of moral luck becomes striking in this context and has, therefore, been frequently discussed (Grinbaum and Groves 2013; Stilgoe et al. 2013). Since innovating is an activity performed in societies by individuals affiliated to companies, universities, or being otherwise tied to collective bodies, chapter six scrutinizes whether this embeddedness undermines individual responsibility. I will consider arguments for the attribution of responsibility to collective and corporative bodies. In chapter seven a constructive theory of innovators' responsibility will be developed. This virtue ethical framework provides a balanced viewpoint in pointing out some admirable and some vicious traits more frequent amongst innovators. I will emphasize the role of *creativity* as an intellectual excellence and describe the *eagerness* to change the present as an excellence of character that motivates individuals' attempts for improvement (Gardner 1993; Martin 2007) Two cases will exemplify this analysis in order to provide *normative orientation*. Chapter eight summarizes the most important insights of this book and offers an outlook for future research.

2. A Humanist Ethics of Innovation

2.1 Innovation

Politically en vogue, an economic buzzword, a scientific puzzle, a subject of psychology, the center of art, a problem for responsibility: The concept of innovation and accompanied notions increasingly beset debates in different areas of science and society (Nowotny 1995). The concept of innovation is thriving. The present enquiry assumes that visions or narratives of the future in general, are expressions of and can be motivators for innovation processes. As Frank Geels and Wim Smit argue: "[…] future images may be intended as interventions to affect the direction and speed of technological developments." (Geels and Smit 2000, 879) Visions' use, misuse, or the omission to create and use visions—their performativity—provide a starting point to discuss issues concerning responsibility in the innovation process and to develop an ethics of innovation (Michael 2000, 33). In this opening chapter, I will scrutinize the concept of innovation at first.

In the following, I will argue that the main addressees of responsibility are human agents. The ethics of innovation developed in this book is considered as a humanist ethics, as opposed to structural approaches to responsibility such as RRI and a recently proposed postphenomenological approach of moralizing technology (Verbeek 2011). A humanist ethics means that people are responsible for who they are and what they do—two moral dimensions that are closely connected (Frankena 1973, 65). In this book, I will focus on the morality of human agents with regard to technological developments. Thus, I will argue that there cannot be an intrinsic value in innovation structures (although it seems that certain structural features contribute to the success or failure of innovations), nor is it reasonable to attribute moral values to "human-technology-associations," as postphenomenologist Peter-Paul Verbeek suggests. A normative theory of innovation should justify *moral standards* for the responsible development of new technologies that relevant agents have to meet. These ideas will be set out in more detail in this and the following chapters. Typically, when scholars from TA (which is also the institutional background of the present research) and Science and Technology Studies (STS) are concerned with innovations, they focus on technological innovations.

© Springer Fachmedien Wiesbaden GmbH, part of Springer Nature 2018
M. Sand, *Futures, Visions, and Responsibility*, Technikzukünfte, Wissenschaft
und Gesellschaft / Futures of Technology, Science and Society,
https://doi.org/10.1007/978-3-658-22684-8_2

However, innovations are not necessarily technological. A proof of Goldbach's conjecture, for example, must be considered a true innovation in mathematics. Only when technology is understood in a very broad sense, can a novelty be understood as a technological innovation. The arguments developed in the following apply to all kinds of innovation, be it social, cultural, mathematical, or technological. Picasso's cubism and Martha Graham's modern dance belong as much to the category of innovations as Einstein's theory of relativity (Gardner 1993). Hence, Homer Barnett's classical definition of innovation articulates my own broad understanding of the concept. The following quote from his book, *Innovation: The Basis for Cultural Change* provides a good starting point for the discussion:

> An innovation is here defined as any thought, behavior, or thing that is new because it is qualitatively different from existing forms. Strictly speaking, every innovation is an idea, or a constellation of ideas; but some innovations by their nature must remain mental organizations only, whereas others may be given overt and tangible expressions. "Innovation" is therefore a comprehensive term covering all kinds of mental constructs, whether they can be given sensible representation or not. A novelty is understood in the same way; hence, "innovation" and "novelty" are hereafter used synonymously [...]. (Barnett 1953, 7)

In a book contribution titled Technological Visions as Subject for an Ethics of Innovation from 2016, I argued that changes in artifacts, thoughts, or behavior are considered innovative compared to other such entities regarding certain parameters (Sand 2016, 334). I considered the adjective "innovative" as a three-digit predicate: A is innovative compared to B regarding a certain feature x. For example, cars nowadays are certainly more innovative regarding their fuel usage and related power production compared to vehicles from the 1950s, which are slower and utilize gas less efficiently. Yet in other respects—for instance, regarding their proneness to failure—they are less innovative and their functionality declined. Through this line of reasoning, innovation has been interpreted analogously to enhancement as a three-digit relational predicate (Grunwald 2008, 250). When we compare enhancing and optimizing as practices, we realize that optimizing as much as perfecting are practices towards a final goal, the perfect or optimal state. In contrast, enhancing is a potentially boundless process that necessarily follows a certain direction; say, for instance, enhancing the environmental footprint by reducing carbon dioxide emission. In contrast to annihilating humans' environmental footprint, which implies a clear goal, enhancing the footprint is a potentially open-ended endeavor. Innovating, as I understood it, can be seen in an analogous fashion as a boundless process in a certain direction. However, focusing like this on "innovative" as an adjective is misleading. Nowadays, innovativeness has a distinctively positive connotation that is mistakenly transferred into the above kind of reasoning. In the example of the car, this becomes clear: One could simply replace the term "innovative" with "better" and thereby maintain synonymy between both sentences. The adjectival use of innovation is clearly normatively laden: Companies are *required* to be "innovative" which suggests that there is an inherent value in being innovative, or that being innovative is considered an important factor for becoming a successful economic player (Godin 2015, 7). As Helga Nowotny writes: "The need to be constantly innovative is felt throughout the corporate world of enterprises that presently undergo another wave of structural adjustement and realignement in the wake of increasing international competition." (Nowotny 1995, 2) The demand for innovation can thus be considered as being part of the current Zeitgeist (Banse 2012, 41).

Innovation in its current adjectival use resembles without a doubt those positive expectations and hopes that were associated with the concept of *progress* until the middle of the twentieth century (Basalla 1988, 132; Gutmann and Weingarten 1998, 11). Developments in science and technology have long been categorically considered as progress. By using this label, such processes were augmented with the notion that technological development necessarily furthers the interests of mankind which must—as common sense suggests—be beneficial or good. Writers from different fields emphasize the link between innovation and progress, while often the positive connotation of the latter is transferred to the former. For instance, in their introduction to the *Handbook of Organizational Change and Innovation,* Andrew van de Ven and Marshall Poole write, "innovation is an important partner to change. It is the wellspring of social and economic progress, and both a product and a facilitator of the free exchange of ideas that is the lifeblood of progress." (van de Ven, Andrew and Poole 2004, xi) Here, innovation is clearly seen as an important condition for progress and both are considered as good (in themselves). Since the twentieth century, the term innovation has been increasingly used to describe processes in science and technology (Zingerle 1976, 511). In contrast to the currently widespread understanding of innovation as scientific or technological progress, Francis Bacon understands the concept in a wider sense in accordance with Barnett's definition to which I am also more appealed. In his essay, *On Innovation* which he wrote in 1625, he mentions the reformation of states and their political systems as an example of innovation (Bacon 1985, 132). He is skeptical about such reforms and warns his readers "not to try experiments in states, except the necessity be urgent or the utility evident; and well to beware that it be the reformation that draweth on the change, and not the desire of change that pretendeth the reformation." (Bacon 1985, 132–133)[2] At the end of the passage, Bacon rejects attempts for reformation that are solely based on the desire for change, when there is a lack of necessity and evident utility. In these attempts, it finally appears that the innovation is more like a stranger, writes Bacon, "more admired and less favoured." (p. 132) Bacon's critical viewpoint stands in stark contrast to the predominant understanding of innovation nowadays. It seems—as indicated in the quote of van de Ven and Poole—that the current debate on innovation is largely lacking skepticism about the value and meaning of innovation, and furthermore focusses on the scientific and technological context as the primary area for innovation.

Hence, we should raise the question whether the noun "innovation" carries the same positive connotation as its current adjectival use and its familiar counterpart—the concept of progress—suggest. I assert, just like Barnett and Bacon, that innovation is plainly to be equated with novelty and as such is neither inherently good nor bad. The identification of innovation with novelty is etymologically obvious. The stem of the term "innovation" originates from the Latin term "novatio"

---

2   Bacon's essay and particularly the passage that critically deals with the naïve trust in innovation should be taken into account for a proper understanding of his view on technoscientific development. His utopia *New Atlantis* is usually understood as a call to "nurture investigation in the natural sciences." (Bruce 2008, xxxv) On the one hand, Bacon argues that "for surely ever medicine is an innovation" that should be taken if one wants to avoid new evils (Bacon 1985, 132). On the other hand, in light of the earlier quoted passage that conveyed Bacon's skepticism, Susan Bruce' reading of *New Atlantis*—which is mainly a critical narrative about the relationship between knowledge and political power—develops plausibility.

(Zingerle 1976).[3] Innovation is an ambivalent process. Its ambivalence may originate from *varying evaluations* of the respective and inevitable effects of change (Ropohl 1991, 250). The qualitative differences in the existing forms of innovation that Barnett mentions can be both positive and disastrous. Any innovation—even one that is less tangible and overt as new artifacts—forces one to adapt or adjust to the novel in ways that can be considered good or bad. This idea of ambivalence must be clearly distinguished from a naïve subjectivism. The diverging and possibly contradictory evaluations of innovations provide no reasons to doubt that normative assessments of new technologies are ever objective (Martin 2007, xiv). Consider, for instance, an innovation in the automation of car production, which can result in a decline in physical labor. Although this aspect of automation is usually welcomed, it also means a reduced appreciation of craftsmanship and possible unemployment for skilled workers. Note that the decline in the demand for physical labor is strictly speaking not a side effect of the innovation. A recession of physical labor is one of the *purposes* of this novelty. Overall, such an innovation might further the company's competitiveness. Hence, while the company might reasonably welcome such an innovation in automation, the process is certainly ambivalent (Nye 2006, 109–134; Grunwald 2009, 159).

Furthermore, even when certain novelties are generally welcomed, they often result in other undesirable changes, which might be considered as side effects. In his analysis of the consequences of philosophical innovations (such as Descartes' project of pure enquiry), Rainer Specht argues that "the development of new problems in the context of those solved is an important stimulus for progress, because it provokes further attempts for solving. Speaking properly of innovations must include mentioning its resulting burdens." (Specht 1972, 15, 211)[4] Whether or not newly developed problems are worse than the ones solved, the innovation requires thorough evaluation. Often the innovation overcomes certain drawbacks and fulfills preexisting desires and, by doing so, it raises further ones. Through the metaphor of the stranger mentioned earlier, Bacon also makes us aware of the possible

---

3 Therefore, Bacon's short essay On Innovation is rightly translated into German as Über Neuerungen (Bacon 2011).

4 Own translation

insecurity resulting from an altered pattern, or a novel thing or idea. However, Specht's conclusions are more pessimistic than Bacon's and they deserve a closer look. Reflecting on the exuberantly positive picture of innovation that currently prevails, he expresses his worries concerning the lack of realism held by people who, in his opinion, orient themselves solely towards the future (Specht 1972, 211). This lack of realism is supposedly a result of permanently dealing with the not-yet-existing, the typical mode of thinking about the future. He evaluates previous religious eschatologists as rather modest compared to modern (scientific and technological) progressive thinkers and innovators (ibid., pp. 225–226). The innovator seeks improvement by renovation or reformation. Such an attitude requires focusing on what is lying ahead. Thus, Specht assumes that through this exuberant focus on the future, a stance towards the present as an intolerable state is reinforced. The present is treated like an ill patient that ought to be cured. The innovator's enthusiasm for the future should be seen as an unconditional will to escape the present (p. 225). Specht's analysis is persuasive in at least two respects. First, as I outlined before, the notion of innovation as being ambivalent is increasingly replaced with a blind trust in the positive values of novelty and change. Thus, many people currently involved in technological development consider novelty as the Holy Grail and the source for good business, social welfare or, indeed, humankind's salvation (consider, for example, the Transhumanist movement). In this regard, the current discussion about innovation clearly resembles the hopes that have long been associated with the idea of progress (Ropohl 1991, 240 ff.). However, the downsides of the technological and scientific developments of the past century are too obvious to maintain a dull trust in the inherently positive nature of progress and innovation. Second, Specht rightly argues for reciprocity between the perception of the present as flawed or imperfect, and pursuing a better future through innovation. He believes that the blind quest to create novelty contributes to the perception of the present as a precarious state. While we can approve this conjunction of future orientation and critique of the present, I will argue in forthcoming chapters that the latter is the prerequisite of the former and not, as Specht suggests, the other way around. The exemplary innovators that I will discuss in later chapters perceive the present as being in need of improvement and, therefore, orient themselves towards a better future that relies on novelty and change.

Following up on Specht's analysis, we should ask: What would be the right stance towards the future, taking the ambivalence thesis and Specht's critique properly into account? He contrasts the progressive—the person who "lives in the future" and strives for improvement through novelty—with the conservative, characterizing the conservative as pursuing stagnancy (p. 211). This confrontation is misleading though, because by attempting to conserve certain aspects of the present, the conservative's focus is also on the future. The conservative and the progressive differ only in their preferred *mode* of progressing through time. One of them is aiming for change, the other one is aiming to conserve, and both of these goals lie temporally ahead of them.[5] A reasonable suggestion would be to take the phlegmatic as the "true" opposite of the progressive in his or her stance towards the future. In contrast to both the conservative and the progressive, the phlegmatic does not pursue *any* goals for the future. Such a person has accepted the present as is—often psychologically through depression and the like—or is dissatisfied with the present but incapable to motivate him- or herself to do anything about it. "Doing" means acting towards a certain goal. One might have wishes about the past—for instance, to have never been born or to have done something different from study philosophy. However, one cannot possibly act towards the past, because goals are always directed towards the future. This type of temporality is essential for the structure of action (Grunwald 2000a, 35).[6] Again, while both the conservative and the progressive aim for a better future in their particular modes and thereby focus on what is lying ahead of them, the phlegmatic fully holds on to

---

5   In his fabulous essay on Conservatism, Ted Honderich smugly remarks about Robert Nozick's political conservatism that "he clearly is no advocate of exactly the status quo." (Honderich 1990, 4)

6   Aristotle writes: "No past event is an object of choice–e.g. nobody chooses to have been the sacker of Ilium–because nobody deliberates about the past either, but only about a possibility in the future; and it is impossible for what has happened not to have happened." (1139b 6–9) This also applies in cases where at first sight the past seems to be the primary "object" of change (Kaiser 2015). The machines in Terminator aim to change the past in order to destroy the human resistance in the future.

the present and does not develop any expectations towards the future (maybe because of a lack of motivation or courage resulting from previous failures). Thus, if the opposite of the progressive is to be found in the phlegmatic, one must ask whether this remains the stance Rainer Specht wants us to take. His reasoning focuses on the flaws of future-oriented attitudes, an argument that, as I have shown, tackles both the conservatives' and the progressives' position. Both inherit a tendency to *discriminate* against the present. Yet, if the ambivalence thesis of innovation is taken seriously, phlegmatism should also not be adopted, because the phlegmatic discriminates against the possible benefits of change and novelty.[7] The chance that the future entails improvements is an aspect of its ambivalent nature. Anyone who clings to the present and accepts its predominant patterns and conditions carries a responsibility for omitting to strike a pathway towards a better future.

In the light of the ambivalence of innovation, the question throughout this book will not be whether we should approach the future at all, but which future should we approach and how? This sheds a preliminary light on the morality of innovators with whom this research is mainly concerned: the ambivalence of innovation makes it unreasonable to blame them as such for their open and expectant attitudes towards the future. In chapter seven, I will refine this verdict and justify valuable assets that support their decision-making and denote these traits as virtues. Furthermore, I will argue for the advantages of such notions in contrast to traditional ideas of responsibility, which rely more on the concepts of accountability and compensation.

In summary: In this book, innovation is understood as a qualitatively different behavior, artifact, or idea compared to existing ones. By discussing Rainer Specht's skepticism about innovation, I have shown that innovation is essentially ambivalent. Furthermore, I have suggested that such ambivalence affects the evaluation of one's stance towards the future. Phlegmatism neglects the possibility of improvement *ex ante,* and conservatism and progressivism are both directed towards the future, yet in different modes. Specht's argument has shifted the focus

---

7   With reference to my reasoning in section 3.6, we could interpret the phlegmatic
     attitude as negligent regarding possible benefits.

from the *process* of innovation to *people's attitudes* towards it. However, who is in control of the innovation process? Is there any place for agency in this process?

2.2   Systems, Evolution, and Agency

What is the origin of novelty? There is a wide variety of candidates that could explain the frequent occurrence of novelty, ranging from metaphysical entities like God, to natural forces, to inherent social dynamics that might be (and have been) considered as roots of novelty and change. Francis Bacon mentions a rather fanciful idea about the main source of innovation. His and other proposals will be critically discussed in this section. Bacon writes:

> For time is the greatest innovator, and if time of course alters things to the worse, and wisdom and counsel shall not alter them to the better, what shall be the end? (Bacon 1985, 132)

Leaving his evaluation of alteration (which leads inevitably "to the worse") aside, this quote unmistakably introduces Bacon's idea that time is the main source for innovation. If we take this statement seriously, we must consider whether our previous reasoning is flawed, because thus far our discussion has focused on the human attitude regarding innovation, as if this was one of the primary sources of its occurrence. In the last paragraph of the previous section, I discussed what the correct stance on innovation as a future-oriented activity was. I also asked whether innovation should be neglected in general. Should we focus on the present and conserve the current patterns, or do the opposite and strive for novelty and change? I argued that both preservation and conservation are future-oriented activities and that one should not *ex ante* neglect the possibility of improvement through innovation. This is, as argued before, the blameworthy attitude of the phlegmatic person. Bacon, however, warns us not to experiment in states (ibid., p. 132). While time, as reflected in the previous quote, might be the greatest innovator, it is not the only factor in Bacon's concept of innovation. *We* must not experiment with states, and thus he suggests people as being capable of bringing about novelty too. We must ask ourselves though, how should we understand Bacon's idea of time as being the greatest innovator? If at all, one might argue, innovations *occur* over the course of time. A novelty follows temporarily from something existing; it cannot precede it. The meanings of "new" and "novel" contradict the meaning of "unprecedented." However, this temporal dimension of innovation does not mean that *time* can reasonably be substantiated as a subject capable of acting towards the future and thereby becoming an innovator. In sharp contrast to Bacon's stands Barnett's definition of innovation. The previous section included his argument that, "strictly speaking, every innovation is an idea or a constellation of ideas." (Barnett 1953, 16) Thus, for Barnett, innovation can only occur through entities that are capable of having ideas. Depending on how wide one is willing to interpret the concept of an idea, things that are capable of having ideas includes living organisms such as humans, and maybe some animals. It is unlikely, that nature or time—as Bacon suggests—can be the bearers and originators of ideas. Thus, if this is innovation's precondition, time cannot be an innovator. Barnett goes on to write:

> Every innovation is a combination of ideas. The only bonds between its parts in a cultural setting are mental connections; they are instituted with the first individual to envisage them, and they dissolve with the last individual mind to retain a recollection of them. The mental content is socially defined; its substance is, in major part, dictated by tradition. But the manner of treating this content, of grasping it, altering it, and reordering it, is inevitably dictated by the potentialities and the liabilities of the machine which does the manipulating; namely, the individual mind. (Barnett 1953, 16)

Barnett focuses on the individual mind in the above quote. He also suggests, however, that what is in the mind of the individual is socially determined. The goals of the individual mind are somehow related to the tradition in which their substance stands. Thus, Barnett sees the human as the main source and driver of innovation, despite being embedded in a specific socio-cultural framework. He encapsulates his view later when asserting that "all cultural changes are initiated by individuals." (Barnett 1953, 39) This idea seems to render the inventor as the genius who is the sole source of novelty. This certainly reflects how history has treated innovators for some time in the past. Such persons were more frequently credited for their inventiveness. One regarded them as libertines that stood against the intellectual habits and customs of their historical periods, thereby bringing novelty and disruption into the world. This has been the perspective many people held about individuals such as James Watt, Thomas Edison, and Alexander Bell (Isaacson

2014). Hans Joas has shown that such mystification of the genius, which also predominantly occurred in German philosophy in the nineteenth century based on the tradition of Herder, resembles the Christian idea of the making the universe from scratch—as a *creatio ex nihilo* (Joas 2012, 111). Despite this interesting historical interference of the creation myth and awe for the inventive genius, the question of whether agency has a place in the innovation process remains unanswered. Many authors agree with Barnett's notion of putting the individual in the center of their concept of innovation. Mathias Gutmann and Michael Weingarten argue that innovation is best reconstructed in the light of a classical theory of action as a multi-digit doing predicate (Gutmann and Weingarten 1998, 15).[8] Thus, someone (a) innovates (b) something (c) according to specific ends (d) using certain means (e) in relation to a certain context (f). The digits (a-f) build the foundation of this reconstruction. In Gutmann's and Weingarten's reconstruction, an innovation is a product of a purposeful (d) activity (b), executed by certain individuals (a). Without any emphasis on the possible divine touch of the agent, the innovation is understood as a *product of human action and will.*

While this notion may seem appealing initially, it carries some difficulties that I will briefly discuss. First, when we consider the development of new technologies or other innovations such as developments in language, art, or science, we realize how little these developments are created from scratch. Rather, they often belong to long-standing traditions and exhibit only minor variations over the course of time. In these branches change is often observably rather incremental than abrupt. This becomes obvious when technologies are considered thoroughly in relation to their antecessors. Contrary to the notion of creating novelty from scratch, we must assume, as historian George Basalla puts it "that even the simplest of artifacts has an antecedent." (Basalla 1988, 55) What does this mean? We can find a stringent continuity in the stages of development of almost all technologies. They do not appear in isolation. By emphasizing the weighty roles of certain

---

8   This is also in accordance with my interpretation of Armin Grunwald's concept of
     enhancement. Grunwald understands enhancing merely as an activity directed to-
     wards certain ends (Grunwald 2008, 250).

ingenious innovators, some historians have decontextualized them and their products from their historical backgrounds and the predominant state of technological development at the time. Thus, many historians have created a bias regarding the disruptiveness of certain artifacts by factoring out the role of preceding artifacts and the socio-cultural backdrop in which they occurred. It is no wonder that it often seems as if technologies like the letterpress or the steam engine came out of nowhere, and rest solely on the shoulders of creative individuals who have advanced the predominant state of technology significantly in a moment's notice. This is apparent in the myth of the development of the steam engine. The fanciful popular account of its development recalls its beginning as the moment James Watt "watched steam rising from the spout of a tea kettle." (Basalla 1988, 35) Under such circumstances one can only consider this as the "lightbulb moment of a lone genius," as this common understanding of having an ingenious epiphany is poetically circumscribed by Walter Isaacson (Isaacson 2014, 479). However, Watt's moment of enlightenment must be considered in light of Thomas Newcomen atmospheric steam engine of a similar guise from 1712. Watt's ideas for improvement came from the "dissatisfaction with a small-scale model of a Newcomen engine he was asked to repair," as Basalla notices (Basalla 1988, 35–37). For many other well-known technological innovations with which we are familiar, almost consistent continuities from their antecessors can be drawn.[9] These considerations lower the plausibility of the *ex nihilo* idea of technological innovation significantly. It also reduces the scope in which individual creativity can be credited with the rise of novelty (Schienstock 2009, 94). Maybe it even *eliminates* individual contribution altogether, which ought to be discussed thoroughly. Before doing so, let me briefly outline the second obstacle that a theory of innovation focussing on individual action faces.

---

9   In his summary of the evolution of the digital age, Isaacson notes: "Therein lies another lesson: the digital age may seem revolutionary, but it was based on expanding the ideas handed down from previous generations. [...] The best innovators were those who understood the trajectory of technological change and took the baton from innovators who preceded them." (Isaacson 2014, 480)

The continuity I described above is closely connected to the apparent fact that innovations are developed in complex social networks and in reciprocal mediation between artifacts and human knowledge. For instance, if we consider large-scale technologies like particle accelerators or power plants, we cannot reasonably suggest that these are products of a single person's inventive mind. It is obvious that in such large-scale technologies, many generations of technological development are consolidated. Moreover, several thousand people work together and merge their theoretical knowledge and technological skills. This is true, of course, for the organization of this process itself, which is not a lonely task either. The knowledge of mechanical engineers, electrical engineers, and physicists is, thereby, itself technologically mediated (Verbeek 2011). Nowadays, throughout their education, research, and daily work, all engineers rely on predominant technologies as sources to gather or analyze data or to facilitate certain tasks. The calculator is a typical example of how individual knowledge—the execution of computer operations—can be objectified in an artifact (Ropohl 1991, 189). The calculator reifies a form of supra-individual knowledge that can be reproduced and made accessible to a large number of people. Anyone who is able to use the calculator can utilize a form of knowledge previously restricted to certain individuals or not available to anyone. For example, few people know how to extract the root of whole numbers. From the possible universal accessibility of this form of computational knowledge and its supra-individual character one can consider this as a form of technologically mediated socialization (ibid., p. 192). The calculator possibly mediates knowledge across large populations, a task that is usually executed by pedagogic institutions and parental education. In the computer as a successor of the calculator, it becomes obvious that such generalized and objectified knowledge outperforms human intellectual capacities in several respects. The analysis of large amounts of data required by engineering research and development nowadays is unimaginable without the assistance of strong data processing machines. Long before human intellectual capacities were outperformed, their physical capabilities required assistance from artifacts such as the steam engine and cranes. Furthermore, due to increased complexity, managing these devises becomes in itself a particular challenge that requires learning certain skills. Engineers need basic insights into programming to make use of a computer's capabilities (Martin 2007, 11). It makes sense to speak of these interlacements of human knowledge and technological skills as technological systems of action ("technologische Handlungssysteme") (Ropohl 1991, 192). The production of new technologies and innovations relies heavily on such reified knowledge in the shape of previously developed artifacts directly and indirectly. Innovations are inseparably intertwined with preexisting technological patterns—not only as their antecessors (as the continuity thesis suggests)—but as their *mediators*. This is with regard to both individual technological skills and wider systemic socio-technological relations.

Furthermore, regarding the thousands of patented innovations that do not find their way to the market, it must be mentioned that even when novelty is created, the innovations are only rarely commercialized (Basalla 1988, 69). Many of the innovations that are registered as patents will never be industrially produced and widely distributed. The innovations we are familiar with have undergone this step, but there are thousands of forgotten innovations. Therefore, one usually distinguishes roughly between the phase of innovation and the phase of diffusion (Rogers 1995, 35). According to this scheme, the first phase in bringing novelty into the world is the creation of it. In the second phase, the idea or artifact is communicated, reproduced, and finds its way to the market. It will then be generally applied in its specific niche or setting. This picture is, however, extremely simplified. It falsely suggests a one-way stream from innovation to diffusion, a course that moreover idealizes the innovator as the primary source of this development. This process is neither linear, nor a one-way stream (Urry 2016, 75). Often artifacts even change their design throughout the innovation process. For instance, sometimes the original design has to be changed to fit the existing technological infrastructure, or this infrastructure needs to be extended or enhanced (in its most basic sense, this means, for instance, the innovation's adaptability to the existing power supply system—which is currently an obstacle for the commercialization of electric cars). This also applies to the natural environment in which technologies are used (for example, in hilly countries mountain bikes will likely receive more consumers' attention than road bikes). Furthermore, if consumers do not accept the technology or divert it from its intended use, they force the producer to adapt it. In the passage from prototype to industrialized production, the innovation faces the most challenging obstacles. Often artifacts must be altered to be produced on a large scale and thereby maintain a reasonable price together with their functionality. These are just a few aspects of the innovation dynamic. Regarding these conditions, it is more appropriate to speak of the "innovation journey," as Arie Rip does, instead of using the one-sided innovation-diffusion model:

> Innovation journeys do not occur in a vacuum. They are part of larger processes, and are entangled with organizations, other technologies, sector dynamics, and anticipations of, and responses from, society. […] In society, a division of labour has emerged between developers of technology and recipients, and there are framework conditions like the patent system, regulatory measures, insurance, and the role of consume organizations. So it makes sense to speak of the context of innovation journeys, and inquire how this context influences the dynamics of innovations […]. (Rip 2012, 158)

In the above quote, Rip mentions some of the central aspects of the innovation dynamic. Again, it is clear that according to these aspects, the influence of the individual innovator appears to diminish and is almost non-existent. Throughout this book, when I speak of the complexity of innovation processes (as in chapters three and seven), I refer to both the multi-agency involvement and the non-linear course of an innovation's development and diffusion, which are represented in Arie Rip's notion of the innovation journey (Urry 2016, 60). Rip rightly notices that besides the abovementioned aspects (environment, multi-stakeholder involvement, technological infrastructure, and technologically mediated knowledge), certain framework conditions also influence the path of the innovation journey.

These aspects encourage one to look for a different type of innovation theory. Regarding multifactorial developments in complex settings, do these insights support an agent-centered approach to innovation, as Barnett suggests? What would be the alternative? At first sight though, it looks as if non agent-centered approaches to innovation develop more plausibility in light of the findings above. Considering the complexity of the innovation journey, understanding an innovation as the emergent product of an innovation *system* might be a more plausible account than considering it as the result of individual creativity.

Thus, a few authors have attempted to draw an analogy between the innovation journey and the evolution of new species in the natural realm. The idea seems obvious, since novelty does not only occur in human activities, but has apparently been occurring in nature before humans existed. Nature brings about novelty all the time.[10] In evolutionary theory, the source of novelty is random mutations and recombination (Mayr 1979, 16–18). The better the phenotypes resulting from mutations fit their natural environment, the better they reproduce. The natural environment includes the occurrence of predators, the divergent climate, and geographical conditions such as an area's location, the availability of food, and the number of other specimens that share limited natural resources (Mayr 1979, 13). If a certain phenotype assimilates into the environment, it reproduces more successfully and thereby passes the mutation on to the next generation. If the following generation also benefits from the mutation, it will further pass on the mutation. Thus, throughout a number of generations a dominant genotype (with its respective phenotype) establishes and becomes a statistically significant feature of the population or the trademark of a new kind of organism. This process as a result of reproductive success depending on an organism's suitability to its environment, is called natural selection (Mayr 1979, 23). Natural selection or "descent with modification" is a theory that explains the development of species. Thus, George Basalla takes up evolutionary theory and analogizes it with technological development in his book, *The Evolution of Technology*:

> Darwin's theory, therefore, is perfectly compatible with the mechanical kingdom. The history of technology is filled with examples of machines slowly changing over time and replacing older models, of vestigial structures remaining as parts of mechanism long after they had lost their original functions, and of machines engaged in a struggle for survival, albeit with the help of humans. The animal or plant breeder **who practices artificial selection by choosing certain specimens for propagation is doing precisely** what the machine builder and the industrialist

---

10   In contrast, Richard Taylor writes: "And raindrops fall, but that a given raindrop should fall at one time rather than at another does not matter, nor is it even easy to make such temporal distinction. Each drop is just like any other, and no newness is introduced by supposing that it falls earlier, or later, or even millions of years earlier or later. Such a world is without novelty. [...] Nothing, in short, would ever be created in such a world." (Taylor 1987, 676–677) Ambivalently enough, he later adds: "Except for the long and gradual changes wrought by biological evolution, nothing new or different occurs in the world." (Taylor 1987, 677)

do with mechanical life **when they plan a new technological venture** [own emphasis]. (Basalla 1988, 16)

This quote is remarkable in several respects and should be discussed briefly. Contrary to Basalla, I assert that the evolutionary theory of technological change entails at least three problems. The first is that the evolutionary theory itself contains certain flaws that are also heatedly debated in the biological sciences (Gutmann and Weingarten 1998, 15; Grunwald 2000b, 56). As is often the case when Darwin's ideas are transferred to other realms like the development of markets or technology, these difficulties are neglected. Consider the unresolved problem of individual versus group and species selection. What is the primary entity of which evolution supports its survival? Usually proponents of population theories, such as Darwin, have suggested the individual is the primary entity that passes on his genotype, and because of the set of features that belong to his phenotype, the individual is more likely to procreate (Mayr 1979, 37). This idea is most obviously manifested when considering breeding as a selective practice (Toepfer 2013, 47). The breeder chooses the individuals that are phenotypically preferred and prohibits the procreation of other individuals. It will be the genotype of those selected individuals that will be prevalent generations after the first mutations occurred. Practices of cultivation and breeding influenced Darwin's initial understanding of evolution. The "flaws" of certain individuals regarding their own reproductive success might, however, be helpful to support the procreation of other members of a group. This might be the case while the individual's fitness, lifespan, and reproductive success are significantly limited (Toepfer 2013, 48). Take an example that originates from Elliott Sober, which is discussed by Georg Toepfer: when predators hunt for groups of a shorter average height, individuals that are short might benefit from gathering in groups containing taller individuals because the group's average height will be taller. Even the taller individuals benefit from such gatherings because their particular group goes under the predator's radar, although the taller individuals in general reproduce slower (ibid., p. 49). The question is whether evolution actually favors the reproduction of groups, individuals, or species. These entities are not congruent and there is ongoing debate in biology about their status in evolutionary theory. If evolutionary explanations are applied to the innovation context, the problems entailed in such explanations are carried with it. For their application to technological innovation, one would have to set out whether particular artifacts represent individual organisms, and whether it is these organisms or their species (for instance, a "train" or "trains") that are reproduced. What would compose a group with regard to technological artifacts? This question raises further obvious difficulties of analogizing evolutionary theory and technological development.

The transfer of the above-mentioned central concepts of evolutionary theory (selection, fitness, mutation, drift) is often flawed (Gutmann and Weingarten 1998, 16). Consider, for instance, the notion of mutation and recombination that is considered as the source of novelty in the organic realm. What would be the analogue source of novelty regarding technologies? Mutation is assumed to be a random process (Mayr 1979). Note that human creativity as described in Basalla's quote above is the exact opposite of such randomness. He considers machine builders and industrialists as planners of technological ventures. Planning is, however, always a *purposeful* activity. Spontaneous behavior and random events are the exact opposite of the planned process that is Basalla's idea of technological development. John Ziman also recognizes this problem when he writes: "The most obvious difference is that novel artifacts are not generated randomly: they are almost always the products of conscious design." (Ziman 2000, 5) One could try to rescue the evolutionary explanation for novelty by arguing that randomness might not be the source of novelty. Since the actual course of technological development, as the argument might progress, often changes unexpectedly and develops in unforeseeable directions, randomness reoccurs in the process as a lack of predictability (ibid., p. 5). In this regard, it is important to make two things clear. First, a lack of predictability stemming from the complexity of a process should not be conflated with randomness. Randomness opposes nomological regularity.[11] Unpredictability, however, is not the proof that irregularity or randomness is an essential part of a certain process. Unpredictability can also be understood as a lack of epistemic insight. Patterns or processes that initially seem complex and unpredictable might

---

11   Things that are regularly conjoined are not necessarily related as cause and effect. For example, the departure of my train is regularly conjoined with the clock showing 8:40 a.m. However, neither the clock nor the bus influences the other in a causal way. It is, therefore, important that regularity should be understood here as causal or nomological regularity.

not yet be properly understood. Such processes can entail regular, nomological behavior, which is not immediately perceptible. Second, although the feature of randomness in technological development is maintained through this step, the initial place of it—the mutations that bring novelty into evolution—is still disposed of such randomness. Novelty becomes a product of planning and, thus, the evolutionary metaphor is significantly *stretched*. Furthermore, central to evolutionary theory is the concept of species. Species in biology are considered as reproductive classes that do not mate with other species, and share similar genepools as a result. While sharing similar features such as tail length, claws, cardiovascular system, and so on, each individual of a species has a *unique* phenotype and transfers the corresponding genotype to their progeny. Variation and novelty occur in this process, as mentioned before. Hence, evolution is based on sexual reproduction within which the recombination and mutation of genes takes place.

How does reproduction and expansion happen within the realm of artifacts? First, artifacts do not procreate. They do not have sexual organs, nor do they reproduce by mating. They are purposefully and industrially reproduced in smaller or larger scales. What is interesting about this process of industrial reproduction is that artifacts are reproduced according to specific models. A model functions as the prototype *for* reproducing an artifact, and the goal of industrial reproduction is to produce exact copies of the prototype. These reproductions are ideally identical in all respects. Classes of artifacts such as mobile phones or racing bikes consist of numerically disparate objects, which are (ideally) identical in all respects. The prototype of these artifacts provides the model *for* industrial reproduction (Janich 2009, 166). In contrast, scientists who categorize new species in the natural realm take a different "methodical" pathway. By observing sets of unique individuals and prescinding from most of their phenotypical features, they arrive at one or a number of commonly shared features. The feature that does not occur in other organisms that are alike in other respects provides the basis to attribute the status of a "new" or distinguished species. Through the abstraction of certain phenotypical features, one ends up with a "model" of a new species that possesses a specific pattern of, for instance, leaves or spouts. For example, according to models of beaks, individual birds can thus be distinguished and classified according to their species. In conclusion, the transfer of central concepts of evolutionary theory (species, mutation, reproduction, and so on) is flawed, or at least often appears to be forced. The meaning of the theory's central concepts has to be stretched significantly to fit the development of new artifacts. Sexual reproduction has no equivalent in the technological realm and randomness does not, if at all, occur during its "procreation." Artifacts do not have "genes" that are recombined or mutate during procreation (Ziman 2000, 5). The model of species in evolutionary theory does not fulfill the function of being a prototype for reproduction as it does in the realm of technology. Species in technology could be considered either as classes, or a batch of artifacts reproduced according to a specific model based on a prototype. On the contrary, models of species in evolutionary theory are the results of scientific abstraction and classification and are therefore created through a completely different method. They are models *of* the species, not *for* the species. Therefore, the analogy between technological development and natural development is farfetched.

Could there still be value in an evolutionary theory of technological change? Let us recall the merits of such a transfer to come to a conclusion. Many aspects of technological development resemble aspects of biological evolution. The complexity of technological systems and innovation journeys, as Arie Rip described them, makes it seem as if novelty "emerges" out of such a system. Throughout the journey, we cannot locate the origins of the innovation. In this way, the systems view of technological change develops plausibility. Thus, it seems as if other forces besides human creativity and purposeful action shape technological pathways. Furthermore, many artifacts or novelties that are patented will never be widely distributed. This resembles their lack of 'suitability' to preexisting patterns or their "environments" in general. Many artifacts that were once widely distributed have become "extinct"; they are not used or produced anymore. Consider, for instance, the Morse telegraph or the Zeppelin. Additionally, as I mentioned before, like the natural realm, technologies gradually transform over time. Many of these technologies have antecessors that are phenotypically or functionally alike, and can thus be considered their "offspring" that require the same fundamental set-up. Revolutions, understood as radical breaks in the genesis of novel artifacts, are rare.[12] While there is historical continuity in technological development, there are

---

12   An opposing view is defended by (Hughes 2006, 28).

also concurrently wide varieties amongst different kinds of technologies. These
varieties often fit into a specific niche, as is the case in the animal kingdom. There
are hammers that are better for chiseling stone, while others are better for nailing.
Similarly, some bird species are predators because of their claws and beaks; others
have a better makeup for sucking flower nectar. These types are distinguished as
species. Considering these similarities, while at the same time acknowledging the
vast differences between natural and technological development, George Basalla
writes:

> The fundamental differences between the two "trees" [the tree of life and the "ar-
> tifactual tree"] show that we must not forcibly extend every element of the bio-
> logical species into the realm of technology. We may fairly assert that technolog-
> ical novelties are selected for replication, and we need not establish which of these
> new kinds of things are to be designated as a distinct "species" or "type." I there-
> fore use the evolutionary analogy because of its metaphorical and heuristic power
> and caution against any literal applications, not the least, the process of speciation.
> (Basalla 1988, 138)

Despite all their differences, Basalla asserts that there is a heuristic benefit in applying the evolutionary metaphor to technological change. While it is hard to argue
against the heuristic usage of the evolutionary metaphor, it is important to mention
that this strategic move also greatly degrades the *explanatory value* of the theory.
Conclusively, by excluding essential concepts of evolutionary theory such as sexual reproduction, randomness through mutation, and species in applying the theory
to technological change, its explanatory benefits almost disappear. This becomes
especially obvious when considering how many systems could be described using
a number of randomly picked features of evolutionary theory. For example, one
could speak about the emergence of solar systems, scientific theories, or moun-
tainous areas. Solar systems develop gradually over time. Several stars and planets
vanish because they collide with other planets or are soaked up by dark matter. In
this context, one could reasonably speak of a lack of suitability to their environment. Other planets survive over long periods and mutate into different shapes;
they grow or decline. Thus, the evolutionary theory might as well be applied as a
heuristic to describe astronomic systems (Mayr 1979, 92–94). Evolutionary theory
thereby becomes applicable to any kind of process. It becomes an infallible theory

that loses its explanatory value (Grunwald 2000b, 56–57). One should be careful of this major effect when widely applying evolutionary vocabulary.

2.3   Technological Determinism and Postphenomenology

There is an underlying theme in the previous considerations that has yet not been discussed thoroughly enough: technological determinism and agency. To address this, I will briefly discuss Peter-Paul Verbeek's postphenomenological way of moralizing technology. I have mentioned several times in this chapter that technologies do not appear to be the products of individual agents and their creative efforts, but of complex systemic processes. From a certain perspective, innovations *evolve*, as the previously discussed evolutionary theories of technological development suggest. How can there be agency in this process? First, the proponents of evolutionary theory, Basalla and Ziman who highlight the similarities between the natural realm and technology, argue that they essentially differ with regard to their causes or drivers. While evolution is considered a random process, Ziman argues that in contrast, technologies are purposefully developed. They are the result of conscious design. However, with regard to the regular patterns of technological development and their gradual emergence, it would be reasonable to consider technological change as endogenously driven. Thus, like in the natural realm in which species emerge according to their suitability to their environment over time, technologies are adopted and used in particular niches and neglected in others. However, it is incorrect to assume that the behavior of individual organisms is determined according to their evolutionary origins. Evolutionary theory explains the prevalence of certain genotypes and their emergence over time. It does not aim to explain the behavior of individual organisms that are part of these species. Proponents of evolutionary theory assume that there are cumulative effects of individual reproduction according to the suitability of the offspring to their environment and the resulting (statistical) changes in the genotype of a population over time (Toepfer 2013, 43). The theory assumes that selection is the force that influences this drift, the change in the prevalence of a certain genotype. Still, this is not clearly a causal force and it needs to be carefully distinguished from a deterministic understanding of this process. Determinism is the idea that every state of the world is a necessary successor of any previous state. If one had full insight into the laws that connect these states, one would be able to draw a comprehensive picture of the future (Keil 2007). The evolutionary model of biological development does not provide such a perspective. The relation between selection (an organism's suitability to its environment) and its favored genotypes is merely statistical and not causal. Knowing the mechanisms of evolution is an entirely different story than predicting the development of species (Toulmin 1961, 24–27; Grunwald 2000b, 58). Randomness through mutation and recombination plays a major role in evolutionary explanations. Both expose the non-deterministic nature of evolutionary processes. However, the complexity of the innovation journey has led many to believe that there are "laws" that govern the development of technologies. As a result, the process has been described as being essentially deterministic (Jasanoff 2016, 14).

There are, however, two types of technological determinism that are often mixed up and should be distinguished clearly (Ropohl 1991, 193). Both can be shown to be falsely interpreted as being more intrusive or powerful than they really are. The first type of determinism is the one just described. It assumes that the process of technological development is in itself determined, endogenously driven, and unfolds according to laws like the laws of nature. We are acquainted with some of these "laws," as Basalla argues. He assumes, for instance, that there is an "intimate connection" between military needs and civilian aspects of modern industry (Basalla 1988, 158). He writes that "military demand for large quantities of standardized clothing, foodstuff, and weaponry did prefigure, in a sense, the creation of mass markets fed by mass production." (Basalla 1988, 159) Basalla understands military considerations as a selective force for technologies and concludes a discussion of several examples with the following statement: "The extraordinary role played by the military in determining technological choices makes our age unique in the history of technology." (ibid., p. 159) Basalla is certainly right in acknowledging the strong relations between technological development and military considerations. Such patterns certainly exist in technological development. Economic circumstances, for instance, have a regular impact on the development of new technologies. An economic recession will very likely have a hindering impact on the innovation process. Such regularities are, however, misunderstood when considered as laws. The concept of a law—particularly the meaning of a "law of nature"—suggests that certain events are not only regularly but necessarily conjoined with each other (Hampe 2007, 44). Their unity is one of cause and effect and necessity binds them together infinitely. To show that certain technological processes are determined means showing that there is not only a temporal, accidental connection between two events, but also that the events always occur together. This cannot be substantiated with examples of the *temporary* conjecture of certain events like military decisions and technological advancements. The term "technological momentum" is preferable to denote the regularities that occur in innovation journeys and the *rigidity* that some established technologies exhibit and impose on societies. Technological momentum is a more suitable term to describe the (temporarily and locally) limited regularities that can hamper innovation journeys, but which can also be utilized to partially govern and influence them (Nye 2006, 55).[13] I will return to the concept of technological momentum in the next section. In summary, the existence of patterns and regularities in technological development do not suffice to regard this process as being determined.

There is another understanding of technological determinism discussed in the philosophy of technology literature (Ropohl 1991, 193). In the previous section, the close interconnection between human knowledge, action, and technological socialization was described. The calculator was discussed as an example of the objectification or reification of human knowledge. By making this knowledge accessible to potential users, one can speak of a process of technologically mediated socialization. Its usage allows participation in reified mathematical knowledge. These interconnections have been interpreted as determinants of human action.

---

13   We find similar ideas in the discussion on "path dependency." Gerd Schienstock writes, for example: "The development of a new path des not occur as a sudden break from the old one. On the one hand, the development of a new techno-organisational paradigm and its transformation into a new national path takes time and the creatin of a new path in its earlier stage remains often mre or less unrecognized; it cannot challenge the traditional paths in any way. On the other hand, old sectors, although they are likely to shrink, reducigin also the influence of traditional paths in the economy, will hardly disappear in a short period. Instead, they will continue to develop; but they may integrate some knowledge, technologies, organisation forms and institutional structures from the emerging new path." (Schienstock 2009, 94)

Highlighting the close interconnection between human action and technologies, Peter-Paul Verbeek has put forward arguments to justify that technologies are not mere instruments of human purposes. Emphasizing their mediating roles in decision-making processes, he argues that human-technology assemblies should be regarded as forms of agency. He pursues a new type of technological ethics with crucial impacts on an ethics of technological development. His viewpoint deserves a closer look.

After discussing several examples that include genetic testing and ultrasound, Verbeek argues that "technologies contribute actively to how humans do ethics." (Verbeek 2011, 5) He assesses genetic diagnostic tests for breast cancer as a technology that enables physicians to determine the likelihood of developing cancerous cells. The technology, therefore, "organizes a situation of choice." (ibid., p. 5) People who are confronted with the probability of contracting a certain disease have a choice about how to react to these findings. They can ignore the probability of contracting a disease, or choose preventative measures to lower the risk. In Verbeek's words: "The very fact that this technology [genetic testing] makes it possible to know that it is very likely that a person will become ill, [...] makes this person responsible for his or her own disease." (p. 5) Technologies widen the range of decisions people have to make and confront them with choices they have not faced before. The same applies to ultrasound: Ultrasound can reveal a variety of defects or diseases in a child. Parents can react to these finding in certain ways. Overall, such technology changes the relationship between the parents and their baby. Expectant parents may start to personalize the unborn baby and experience them not merely as a growing organism but as their future daughter or son. "Ultrasound constitutes the unborn in a very specific way," Verbeek argues (p. 24). In general, this is how technologies "actively contribute to the moral decisions human beings make," (p. 5) a statement that is very similar to the one I quoted at the beginning of this section. The same applies to technologies that are less morally "questionable." According to Verbeek, technologies like the thermometer shape the ways people perceive reality (p. 9). In general, one can argue that technologies mediate the ways in which humans interact with each other (moral dimension) and the ways in which they interact with the world (epistemic dimension). According to these considerations, Verbeek asserts that the traditional (humanistic) metaphysic of morality is misplaced when facing the reality in which large parts of human actions and perceptions are mediated through technology. Regarding technologies as merely passive instruments of human action is a gross devaluation of their essential roles in the human social life and its interaction with the environment. Such interpretation underrates technologies' ethical significance, which Verbeek considers the "humanist bias." By acknowledging technologies' major contribution, Verbeek aims to establish a new form of moralizing technology that renders them as integral parts of the moral community, instead of neutral artifacts subdued to the human will (p. 42). To grasp the radical notion of Verbeek's postphenomenological ethics of technology, it is worthwhile to consider the following longer quote:

> From a mediation perspective, technologies should not primarily be approached as invasive powers in need of ethical limits but as morally significant entities that need to be assessed in terms of the quality of their impact on human existence. Technologies help to shape human actions and decisions by mediating our interpretations of the world and the practices we are involved in. Therefore, they play a significant role in human morality. Approaching technologies as morally relevant entities has important implications for our understanding of central ethical notions like moral agency and responsibility. By approaching agency and responsibility as phenomena that are distributed among human beings and nonhuman entities, ethical theory can do justice to the hybrid character that many actions and practices have acquired. In fact, the humanist bias that has come to characterize ethical theory needs to be abandoned if theory is to take responsibility for the close connections between technological artifacts on the one hand and human actions and decision on the other. (Verbeek 2011, 153–154)

Verbeek suggests that one has to do justice to the intimate relationship between humans and technologies. According to him, agency should be distributed among human beings and nonhuman entities to do justice to their interwoven relationship and hybrid characters. Commonsensically, one would say that inanimate objects cannot be considered agents. They lack intentionality and control over their behavior. Therefore, one cannot reasonably assume that, for instance, a pair of glasses intends to make its bearer see well. Verbeek, however, does not accept this reasoning. Intentionality, which would exclude nonhuman entities from the moral realm, should not be understood as a necessary condition for moral agency. Technologies are *effective* and should, therefore, be moralized:

> Yet the argument that things do not possess intentionality and cannot be held responsible for their "actions" does not justify the conclusion that things cannot be part of the moral community. For even though they don't do this intentionally, things do mediate the moral actions and decisions of human beings, and as such they provide "material answers" to the moral questions of how to act. Excluding things from the moral community would require ignoring their role in answering moral questions—however different the medium and origins of their answers may be form those provided by human beings. (Verbeek 2011, 42)

When we consider Verbeek's constructive approach of an ethics of technology, this reasoning appears to be less radical than the quote might suggest. A consequence of attributing agency to artifacts would obviously be to not hold them accountable for certain events. If responsibility is accompanied by certain reactive attitudes and moral reactions such as praise, blame, or even legal punishment, one might wonder what the proper moral reactions to the "actions" of artifacts would be. Consider a car tire has a puncture, which causes an accident. Will the tire be blamed or sentenced to jail? Obviously, this would be an absurd reaction. Verbeek wholeheartedly concurs with this reasoning when he writes that "it does not make sense to consider technologies full-fledged moral agents in the way human beings are moral agents." (p. 108) Hence, who (or what) is responsible for the moral design of technological artifacts, for their appropriate use, and their social effects? Regarding his own ethical approach, Verbeek maneuvers along the lines of those very typical suggestions we usually encounter when normativity enters the debate of technological development. Instead of taking the radical interpretation of his notion of "being part of the moral community" into account, he outlines some classic ideas of responsible techno-development. He proposes, for instance, stakeholder analysis as a method of laying bare all relevant arguments for or against certain technologies, and enabling the possibility of "weighing all these arguments against each other." (p. 106) Furthermore, he suggests that relevant actors should be included in the process of technological development, which is introduced through the idea of Constructive Technology Assessment (CTA). CTA "could dispel the fear of technocracy [...] and open a space for deliberative democracy in processes of technology design." (p. 112) Finally, he discusses moral inscription. According to him, designers can embed "scripts" into the design of technologies. These scripts have behavior-influencing effects on users of the technologies. At the beginning of his book, Verbeek mentions a famous example of Langdon Winner: the low-hanging overpasses on Long Island. These overpasses were "deliberately" built so low that buses could not pass and, therefore, could not access the beach. This had the consequence that African Americans, who could not afford cars, were denied access to the beach (p. 5). The architecture of the bridge was exclusive and discriminatory. Verbeek argues that technologies have to be built differently, in order to be less discriminatory and more ethical. Designs should be value-sensitive and produced only after inventors have critically anticipated possible interactions (abuse and tragedies of the commons) with their users. Many of Verbeek's suggestions hold great value for creating worthwhile technologies. However, they do not exceed the humanist idea of an ethics of technology. His prescriptive suggestions address human agents and take into account their capacities for shaping technological development at different stages, as well as their roles as originators of technologies and their respective designs. Thus, these ideas concur with classic approaches to the responsibilities of technological developments (Lenk 1993b, 116). In these works, such as Hans Lenk's classic approaches to the responsibility of engineers, the notion of an anticipatory responsibility was already introduced (ibid., p. 116). Verbeek certainly adds a number of valuable methods and dimensions to these traditional types of responsibility. For instance, he highlights the (behavior-influencing) side effects of technological design and outlines the benefits of CTA, stakeholder analysis, moral inscription, and other methods. However, these dimensions are still covered when considering technologies not as parts of "the moral community," but as mere *instruments* of human purposes. Thus, "instrumentalism" is a perspective that could very well be adopted. Instrumentalism regards agents as the main sources and drivers of technological change. Technologies affect humans in numerous ways. These effects vary in predictability. Technologies can thus be regarded as instruments of human purposes without taking them as constitutive parts of technologically mediated forms of agency. This notion would be perfectly in accordance with a humanist ethics. However, Verbeek criticizes instrumentalism for failing "to take into account how moral actions and decisions are thoroughly technologically mediated." (Verbeek 2011, 88) As mentioned before, he suggests that "technologies cannot have moral agency in themselves," but that technological mediation of moral actions and decisions "needs to be seen as a form of agency itself." (p. 61) Based on these considerations, I will provide a three-stage critique of Verbeek's theory.

First, when criticizing instrumentalism for not taking the mediating role of technologies properly into account, Verbeek begs the question of technologies nature. From an instrumentalist perspective, one can reasonably accept the notion of the behavior influencing, and in general, fundamental role of technologies in human decision-making. Considering them as mere instruments of human purposes often even *presumes* their role as behavior-influencing devices. Thus, street signs are built to inform and adjust people's driving behavior. Their mediating role is part of their design, and their purpose as tools is to produce these effects. Clearly, many of these effects—the ways in which humans interact with technologies, how they might use or abuse them, and the numerous effects the systemic, widespread use of technologies produce—are often not clear or predictable (an issue I will discuss in chapters three and seven). This, however, is not a problem that can be solved by attributing agency to human-technology assemblies. Clearly, mediating and behavior-influencing effects of technologies are not wholly denied by instrumentalists. If the result of developing the ultrasound diagnosis of fetuses is parents' altered perceptions of their offspring, then the instrumentalist can decide whether to change the design of the instrument, accept this as an unproblematic side effect, or even approve it for fulfilling the exact purpose of its production. Such possibilities of technological impact on human perception do not undermine the prerogative of human responsibility to react to these possible outcomes. The need for extended emphasis on the mediating role of technologies, interpreting them as special forms of agency, cannot be brought forward as an argument against instrumentalism, because it is exactly what is in question here: Whether technologies are mere instruments or constituents of agency. Providing an answer in favor of new mediate forms of human-technology agents begs the question of whether technologies nature is active or passive.

Second, as mentioned before, the advantages of emphasizing the interwoven character of technologies, human thinking, and behaving, as well as postulating new forms of "mediated agency" to gain a better notion of an ethics of technology are not clear. Any of the proposed prescriptive or normative assumptions of Verbeek (inscribing morality, using CTA, and stakeholder analysis) are addressed to human agents. These suggestions can be brought forward from an instrumentalist point of view, too. These guidelines do not require in any essential sense the emphasis on the interwoven character between human decision-making and technology emphasized by Verbeek. These considerations can very well be subsumed into a humanist theory of morally shaping technological development.

Third, it is not clear what is meant by "being part of the moral community." This notion is introduced by Verbeek to challenge the thesis that only things that are capable of intentionality can bear responsibility (Verbeek 2005, 216). By providing "material answers" to moral questions, he assumes that artifacts can become part of the moral community (Verbeek 2011, 42). If there is no essential difference between Verbeek's constructive approach to moralizing technology and the humanist approach of moralizing technology, what would justify attributing human-technology assemblies the status of moral agents? By taking a closer look at the variety of human-technology interdependencies investigated by him, it becomes obvious that his reasoning is rather vague. Verbeek mingles passive, active, and mere instrumental predicates when describing the role of technologies on human action and decision-making. Thereby, by attributing verbs to artifacts, he stumbles over his own vocabulary in many parts of his analysis. He argues in the quotes above that, "things do mediate," "they provide material answers," they "constitute the unborn in a specific way," and they "actively contribute to human decision-making." These predicates raise the idea that these entities possess agential powers. However, if technologies have this active role and if they provide "material answers," what are they *aiming* to achieve in doing so and how could they have understood the question they are answering? These two questions make explicit that any meaningful use of verbs is necessarily connected to the ascription of intentionality and the capability to act accordingly. Many of the verbs applied by Verbeek can only be understood in a metaphorical sense, because their literal use presumes intentionality of a certain kind. Consider, for instance, a brick falling on the reigning president, causing her death. This occurrence "answers" the question—to use Verbeek's terms—whether she will be re-elected. However, this usage must be understood as merely metaphorical. If human intentionality has not anteceded this causal chain (someone dropped the brick on purpose), we have a clear instance of an accident. The flat tire that causes the car accident is not to be blamed either. This is an accident, too. Therefore, one must carefully distinguish between killing, manslaughter, and accidents. These concepts carry fundamentally different normative meanings. Each of these nouns denotes events that result in the death of a person. However, a falling brick resulting in the death of the president is merely an accident. Describing the same instance as the result of an action, which is a purposeful behavior, changes the very nature of the whole event insofar as moral reactions like praise and blame become required (Davidson 1980b, 46). Intentional behavior causing the death of a person constitutes a killing (even if it is merely the result of negligence, which one could have been aware of beforehand), or even murder (if it had been planned beforehand). Hence, if a car has been checked carefully and a tire still bursts, the resulting accident is the outcome of a malfunctioning technology, an instrument to get from A to B. It becomes obvious that many of the terms that are preferred by Verbeek to describe the influence of technologies on human decision-making and behavior are ambiguous. "Contributing," "affecting," and "influencing" all have active and passive meanings. They refer to causal potency, but only active or intentional causal potency is morally relevant (I will explore this in more detail in chapters three and four). None of Verbeek's examples shows convincingly that there is any active technological contribution to human behavior or decision-making.

So, how can we describe technologies' roles, if not as active ones? There is no doubt that technologies affect human behavior, thinking, and decision-making. Clearly, technologies add alternatives to the predominant scope of human action. Technologies allow people to print books in large amounts, to know the results of complicated computations, to rescue prematurely born infants through medical means, to travel in a short time across several continents. These opportunities have not existed before; they clearly broaden the range of decisions humans may be required to make and, thus, the range of decisions that need to be morally assessed. In other respects, though, technologies can also narrow the range of alternatives. One cannot simultaneously carry a folding chair on a mountain hike and reduce the backpack's weight to a minimum. Any item one chooses to take on a hike to be more comfortable increases the overall load and limits one's freedom. Yet, the role of technologies in these contexts—their limiting or expanding "power"—is purely *passive*. Any (morally) relevant decision remains up to human agency (which is also at the fore of Verbeek's approach, as shown above). If something's effectiveness in influencing human decision-making of this sort would suffice for something to become part of the moral community, other elements or artifacts that extend beyond technological artifacts need to be included in this community. Consider, for instance, the strong influence natural catastrophes like those that volcanic eruptions or tsunamis have on human (or moral—as Verbeek writes) decision-making.[14] Clearly, these events shape human decision-making. They interfere with previously-made plans and have a crucial impact on how humans behave. They have led people to leave their homes or to make similar morally relevant decisions, such as leaving relatives behind. If effectiveness would suffice for becoming part of the moral community, such events would have to be attributed to moral agency, too. Both natural environments and technologies affect human decision-making. They are closely interwoven with human morality. To put it differently in the words of Verbeek, they (help to) "constitute human lives." Humans and the natural environment are, therefore, to be considered as "mediated forms of agency." It becomes obvious that the semantic *ambiguity* of verbs such as "contributing," "influencing," and "effecting" is the source of Verbeek's misled analysis. He mixes both passive and active meanings of these terms and thereby contributes to the confusion about their normative nature. The interwoven character of human behavior and technology does not provide a profound argument to introduce new types of agencies or other moral entities. If they were, many more than just human-technology assemblies have to be approached as moral entities. Technologies can be seen as *instruments* for human purposes. Sometimes actions that use these instruments have specific, unpredictable, and negative effects on human decision-making. This characteristic of technologies should not be downplayed by any ethics of technology. When technologies "behave" unintendedly like the falling brick from the example above, the usage of the concept of behaving must be regarded as merely metaphorical. Thus, the outcomes of such courses should be considered accidents, thereby comprising a special category of *events* that is different to a category of *actions,* which are purposefully executed.

---

14   A similar argument has been brought forward by Martin Peterson (Selinger et al. 2012, 620).

2.4   A Humanist Ethics of Innovation

Innovations have been characterized as novelties that are distinguished by being qualitatively different to existing patterns, things, ideas, or behavior. The qualitative differences that accompany these novelties are ambivalent insofar as they might solve existing problems or issues, while inevitably causing side effects and forcing one to make adjustments. Innovations must be seen as shifts within a web of existing things, ideas, and behavior that affect the arrangement as a whole. The reconfiguration of existing patterns through novelty is an ambivalent process. It is important emphasize once more the implications of this thesis regarding innovation. Clearly, large parts of the current innovation discourse are biased in two respects. First, the discourse emphatically underlines the positive aspects of innovation, thereby resembling the expectations that have long been primarily associated with progress. Companies seem to believe that they are contributing to their monetary success and competitiveness by defining themselves as innovative (Godin 2015, 7). Governments approach innovations as the ultimate means to face, for instance, social challenges, global health issues, pollution, and climate change. Furthermore, large parts of the scientific system are nowadays dedicated to producing technological innovations, thereby further shifting their focus from "pure" knowledge accumulation to the production of *technological* novelty and realizing technological possibilities (Nordmann 2011, 26). Herein lies the second bias of the current innovation discourse: Most approaches to innovation reduce possible changes and alterations to those currently pursued in the realm of (privately or publicly driven) technoscientific endeavors (Nowotny 2006, 14). Those technological developments have advanced most visibly in the past two centuries. However, innovation must be understood more broadly than as mere technological progress. Qualitative differences and, hence, novelty can occur in the arts, mathematics, abstract thinking, morality, architecture, political thought, culture, and so on. Understanding innovation in this way offers a great advantage. The innovation journey described by Arie Rip is a multi-dimensional process. Novelty is not only an initiator of an innovation journey; it can also influence the journey by qualitatively altering its framework conditions. Civil society organizations and their activities, legal frameworks, patent regulations, governmental policies, purchasing behavior, and other aspects are parts of the complex environment in which the innovation journey takes place. Innovating these framework conditions can also alter the journey significantly. In later stages of this book, I will discuss RRI as a program for structural reform promoted by René von Schomberg that focusses on altering the framework conditions of innovation processes in order to make them more responsible (von Schomberg 2013). Such and similar programs are worthwhile because they emphasize that innovations are *not* identical to technological advancements. These programs for structural reform *presuppose*, however, a crucial element and the focus of this enquiry: the existence of agents capable of reacting to moral demands and of assessing the reasons that speak in favor of, or against, competing possibilities of change.

It is important to repeat that once the value of goals such as challenging climate change, social inequality, global health problems, and other global or local problems is acknowledged, innovations can be proper means to tackle them (Lenk 1994, 41). Helga Nowotny has a particularly profound way of expressing this instrumental value of innovation, writing that "innovation is the only credible response currently available for coping with the uncertainty it has helped to generate." (Nowotny 2006, 14)) In accepting such a thesis—the possible range of improvements through innovation—one has already overcome the innovation-adverse position of the fatalist who, out of a belief in higher forces or a lack of impetus (which would more adequately characterize a phlegmatic), asserts that one cannot change the future and abides by the status quo (Seebaß 1993b, 8). For some time in the middle ages (and in the antics—think of the *fatum* as a notion of the inevitable (Meyer 1999)), this has been a common stance for people believing in an almighty god who is, thus, conceived as the initiator and *guide* of the course of the universe, including one's individual future which eventually results in salvation (Hölscher 1999). Nowadays this viewpoint has lost its plausibility and cannot be a credible perspective. Conceiving the future as a place towards which human purposes and intentions can be reasonably directed and can be significantly altered through human effort, is a specifically modern attitude (ibid., pp. 36–39). With this attitude and the possibility of active interventions into the course of sociotechnical developments, the concept of responsibility emerges, and with it, the typical system of responses to misbehavior or good conduct with which we are all familiar: praise, blame, resentment, awe, and so forth (Strawson 1962). David Nye writes about this "expanding relationship of responsibility" (Mitcham 1987) in the technoscientific realm as follows:

> Until the nineteenth century, people usually considered accidents and disasters, such as explosions, fires, collapsing buildings, or shipwrecks, to be inexplicable "acts of God." Often, people interpreted them as divine punishment. By c. 1900, however, they increasingly demanded scientific explanations. Geologists could show where, why, and to some extend when earthquakes were likely to occur. Engineers could explain how a bridge failed or what component of a steam engine was responsible for a burst boiler. The older moralizing tradition, which saw accidents as God's chastisement, disappeared, and citizens assessed risks, mistakes, and culpability. Formerly inscrutable events became legible to the safety engineer, the tort lawyer, the insurance agent, and the judge. (Nye 2006, 162)

This attitude stands in contrast to the phlegmatic position; it is held both by conservatives who aim to conserve certain aspects of their present (into the future), and those that are more explicitly termed progressives, who seek more radical change. In any case, both have understood the significance of innovations' ambivalence. They acknowledge that innovations can be accompanied by benefits and that it is not reasonable to condemn the process *as such*. Condemning it would mean neglecting the positive impacts on society that innovations might bring. As David Nye and others argue, this typically modern idea relocates the responsibility for the future from an omniscient God or sheer fate to *agents* that have the capacity to reflect and choose from a variety of alternatives (Urry 2016, 94). Having elevated our reasoning about innovation to this point, it seems obvious that the major normative question for innovators, policy makers, and citizens can be stated as: Which innovations should be chosen and how should they to be pursued? John Bessant has expressed this as follows:

> The key issue is around how far we explore and consider innovation in its early stages in terms of the potential impacts it might have, and how far we are able and prepared to modify, ameliorate, or possibly abandon, projects which have the potential for negative effects [...]. (Bessant 2013, 2)

This insight is fundamental for the present enquiry in that the questions raised by Bessant are *normative* (What are relevant "negative effects?") and, therefore, presuppose *agency* to deliberate and determine their answer (Grunwald 1999, 222).

In the previous quote, Bessant asks what can be *done* to make the world a better place and to avoid harm. Which kind of innovations should be pursued? Which innovations should be avoided or abandoned? Or, to put it differently, how should *one* alter present societies? To answer these questions, one has to compare possible pathways towards the future and the means to pursue them. Evaluating such pathways requires certain qualities such as technical and social knowledge, wisdom, eagerness, clear-sightedness, prescience, and practical skills. Such qualities listed here are the excellences of character and intelligence that Aristotle discussed at length in his *Nicomachean Ethics*.[15] Consider a more concrete formulation of a topical and urgent innovation challenge: How should we react to climate change? Should governments encourage consumers to reduce their CO_2 emissions by reducing purchases of certain products, or avoid individualized traffic with ad campaigns and taxes? Should governments offer incentives to companies to develop new means of electrical transport? And if so, what kinds? (Grunwald 2012a; Urry 2016) Should entrepreneurs invest in autonomous driving? Should innovators develop new battery systems, smart grids, or other technological devices that help to combat climate change?

I will not focus on these types of questions in this book and much less attempt to answer them. What is more important to understand is that these concrete questions, as much as the "basic" innovation challenge as formulated by Bessant, confronts *human persons* (innovators, entrepreneurs, policy makers, researchers in TA and STS) with the challenge of decision-making and the execution of plans to successfully reach set goals (Martin 2007, 11). The focus of an ethics of innovation does not lie in an analysis of technological, social, or legal architectures or human-technology assemblies as postulated by Verbeek. Instead, an ethics of innovation focusses on the variety of *people* that *act* within these architectures and how they can utilize these structures as instruments and even transform them. These people

---

15   I will cite Aristotle's Nicomachean Ethics in the present book using the standard Bekker notation. Original quotes are extracted from J. A. K. Thompson's translation of the Nicomachean Ethics which has been revised and published by Hugh Tredennick in 1976 and provided with a new introduction by Jonathan Barnes in 2004 (Aristotle 2004).

have the capacity to intentionally (and unintentionally) shape technological, social, and legal environments. The ethics of innovation developed here, analyzes the basic conditions for responsible agency, taking into account the (often limited) range of possible alternatives agents have, the ways in which they train dispositions and excellences in navigating the maze of innovations to find morally desirable pathways and discover solutions (which constitutes another excellence of intelligence: creativity), and how they harmonize their personal passions and interests with the external demands of morality. These qualities clearly mirror the virtues at the center of Aristotle's ethics. Aristotle's arguments are still topical and substantiate the case for a virtue ethics of innovation. Instead of outlining what innovators, entrepreneurs, policy makers, and other relevant agents concretely ought to do to innovate responsibly; I will present arguments for focusing on the character traits of these agents, which are important conditions to initiate morally desirable change. The arguments presented in chapter seven highlight innovation processes' complexity and opaqueness, which makes finding universal obligations that are applicable to a variety of contexts extremely difficult. Furthermore, the legitimacy of satisfying interests that are neither impartial nor entirely self-centered, like living a meaningful life and fostering social relationships (friendships), is an element that is rarely acknowledged when people's moral responsibilities are discussed. As Mike Martin writes: "[W]e should not neglect the moral importance of scientific understanding itself and the personal commitments of scientists, engineers, and other science-oriented professionals." (Martin 2007, 2) The potency of this virtue ethical framework for responsible innovation is outlined in more detail by discussing the intriguing examples of Steve Jobs and Victor Frankenstein in chapter seven. This analysis will provide ideas of excellence of character and intelligence, as well as vices, both of which can function as orientation for responsible behavior without determining *in concreto* what to do.

This virtue ethical framework is general in nature. It avoids answering the previously mentioned practical questions currently faced by many innovators, entrepreneurs, and policy makers. Instead, it introduces the general preconditions to approach these and similar problems in a responsible manner.[16] Because of its generality, my approach is transferable to contexts outside of the realm of technoscientific innovation. Furthermore, because there are a variety of entry points to innovation journeys, the present humanist ethics of innovation can also provide orientation to different agents capable of redirecting their innovation journey's pathway. As shown in the quote I included from David Nye's *Technology Matters*, the increasing responsibilities of safety engineers, tort lawyers, insurance agents, and judges is emphasized. Other relevant actors in innovation journeys are citizens (not only in their role as consumers, but also as voters and citizens (Grunwald 2012a, 89)) and policy makers such as politicians or representatives of civil society organizations, engineers, innovators, managers, and researchers. It is outside of the scope of this research to deal with the specifics of these professions, their societal roles, and their impact on innovation journeys in detail. Instead, for several reasons, in the next chapter I will focus on the case of "visioneering." Visioneers are an increasingly relevant group of agents in innovation processes, which is why they have received increased attention in STS and TA in recent years (McCray 2013; Sand and Schneider 2017). Visioneers can be characterized as people with scientific backgrounds who foster close ties with tech companies that are increasingly occupying the public discourses in a wide range of emerging technologies such as in vitro meat, synthetic biology, and big data (Grunwald 2015; Dickel and Schrape 2017; Ferrari and Lösch 2017). Visioneering can be understood as a technoscientific practice that enmeshes utopian narratives to propagate visionary technological pathways in public and business contexts with the practical managing, designing, researching, and making of novel technologies (Sand and Schneider 2017). As this type of agent has been attributed a special responsibility for possibly launching technological dynamics by a number of authors from TA and STS, they are worthy of an extended discussion (Simakova and Coenen 2013; Cabrera Trujillo, Laura Yenisa 2014; Ferrari and Marin 2014). The forthcoming analysis will

---

16   A remarkable application of virtue ethics to a number of emerging technoscientific
      fields like surveillance technologies, war robots, and human enhancement is pre-
      sented in Shannon Vallor's Technology and the Virtues: A Philosophical Guide to
      a Future Worth Wanting (Vallor 2016).

introduce the basic conditions for being held responsible. Against these early, initial approaches to visioneering responsibility, I will point out some perils associated with attributing responsibility for the consequences of visioneering activities.

Clearly, agents such as visioneers, innovators, engineers, and policy makers do not act in a social or material vacuum. There are at least three external aspects or contextual conditions that impose constraints on agents and thereby limit their freedom. It is important to understand these contextual conditions not as threats to the problem of genuine freedom (which will be discussed in later stages of this book). Instead, it is preferable to approach these aspects as limiting agents' reasonable *alternatives* (rather than constraining freedom in a metaphysical sense). From a naturalist point of view, genuine or metaphysical freedom appears to be most miraculous type of freedom that imposes a causal, and even ontological, scheme upon actions. Furthermore, the fact that people do not chose their heritage, their early upbringing, and their dispositions—which seem to have an impact on the moral evaluation of persons and their later behavior—is known as the problem of constitutive moral luck. Both of these threats—the naturalist challenge and the problem of moral luck—can be considered as threats *internal* to the agent. I will provide some arguments as to why neither of them undermines the idea of genuine freedom in chapters five and six. The three *external* aspects limiting non-metaphysical freedom are these: a) social and legal environments, b) corporative and collective structures, and c) material and technological environments. None of these aspects undermine agency, either. They provide the framework conditions within which rational decisions about the future have to be made, and such decisions are what people can be held responsible for. Clearly, the material and technological environments impose certain limits on what is possible for agents (Bieri 2013, 46). As mentioned before, the elements and forces of nature set boundaries for what can be done and what is technically feasible. Traveling at the speed of light is physically impossible and, thus, cannot be demanded from humans (Nordmann 2013a). A promise that is broken (for instance, someone's promise to visit a friend living abroad) because of a volcanic eruption is clearly excusable. In a similar manner, already established technological systems are *momentous* which means that they resist simple, immediate, and fundamental transformation. The notion of technological momentum is weaker than technological determinism and, therefore, preferable to pin down technologies' rigid character without suggesting

the inevitability of establishing new technological pathways or substantially altering old ones. David Nye outlines the idea of technological momentum and distinguishes it sharply from technological determinism:

> The concept of technological momentum provides a way to understand how large
> systems exercise "soft determinism" once they are in place. Once a society
> chooses the automobile (rather than the bicycle supplemented by mass transit) as
> its preferred system of urban transportation, it is difficult to undo such a decision.
> The technological momentum of a system is not simply a matter of expense, alt-
> hough the cost of building highways, bicycle lanes, or railroad track is important.
> (Nye 2006, 55)

Similarly, a complex energy system like the one established in Germany, which
relies largely on coal and other fossil fuels, cannot be transformed into a sustainable system that is entirely based on renewables overnight (Urry 2016, 78). The
rigidity of such complex technological systems and the interdependency of its elements make radical shifts in its setup almost impossible. Therefore, demands for
radical and fast changes of these systems directed at relevant actors are unreasonable. This does not mean that these systems cannot be incrementally changed, or
in the case of the German energy system, that many societal agents have not met
their responsibilities and performed possible incremental improvements. However, it is important to keep technological and material limitations in mind. They
affect the realm of possible alternatives. In a similar manner, the social and legal
setups of states constrain what agents are capable of doing and achieving.[17] For
instance, a society that has strong Christian roots will not easily give up observing
Sundays as work-free days including its uncomfortable side-effects (for some citizens), such as closed shops and limited public transport (Grunwald 2000b, 213).
Besides these established social norms, national and transnational legal frameworks impose certain constraints on agents. Unlike technological and material

---

17   Although I emphasize here, and in the following considerations that the social and
      legal setups of states are primarily constraints, it must be clarified that these systems
      are imposed to increase individuals with possibilities of which they are devoid in
      anarchical systems, because they have to live in fear. This is debated in political
      philosophy since Hobbes (Nagel 1991a, 34).

boundaries, legal frameworks and societal norms are "soft regulators." Soft regulators do not strictly undermine the possibility of pursuing alternative paths. Rather, in moments of decision-making and reflection, one has to reckon with the individual and social consequences of breaking laws or social norms. As a result of such reflection it can seem rational to accept legal punishment or social disdain in order to pursue ends whose value outweigh them, especially when the legal framework is illegitimate (Nagel 1991a, 5). Despite the emphasized role as soft regulators, social norms can also be used as facilitators of change. Just like natural mechanisms, social values and norms can be utilized in order to govern innovation (Thaler and Sunstein 2009). Because of the reciprocal interdependency of such norms and technological developments, one can seize these regularities. Take, for example, the social value of individualism which is allegedly central to American consumer culture (Nye 2006, 44). Actors pursuing the transformation of the American energy system could make use of such social value, for instance, by promoting decentralized energy grids as opportunities to increase consumers' autonomy and individuality.

Last but not least, corporate and collective structures demand certain commitments from people involved. Nowadays there are few freelancing researchers without any corporate affiliation (von Schomberg 2013, 53). Private and public bodies that usually employ innovators, visioneers, and researchers have to negotiate their products and research endeavors with a variety of social actors that foster particular expectations towards innovations. Some expect financial revenue (stakeholders), others expect the increasing appeal of voters (politicians), others the maintenance of cultural or historical sights (citizens), or the protection of the environment (non-governmental organizations (NGOs)), and so on. Together these competing interests and expectations constitute a force field in which innovations have to be established. Corporations constitute microcosms of these competing force fields within the wider social system: corporations employ hierarchies and structures that determine who is most powerful in shaping corporative agendas, what the central goals of corporate actions are, and how workflows and deliberative procedures are managed. As I will argue in more detail in chapter six, I do not believe that these characteristics provide good reasons to consider such bodies as agents that carry moral responsibilities. However, these characteristics have to be taken into account when agents who act within these structures are evaluated.

These complex framework conditions of innovation journeys will be considered when dealing with visioneering in the following chapters.

I should reiterate that the outlined features of humans' social, technological, and material environments present no reason to be concerned about their freedom. Neither do these aspects provide grounds to regard these environments' interdependence with human actions as new forms of agency. In contrast, we must see them as the major reasons for the existence of responsibility. If, on the one hand, there were indefinite possibilities, decision-making would be pointless because such a notion implies the existence of an indefinite amount of acceptable possibilities, which makes it hard, if not impossible, to act wrongly. Under these conditions agents would not be praiseworthy because they cannot fail—they would probably not even have to set goals (Bieri 2013, 49–51). If, on the other hand, there were no alternatives at all and agents could not choose to do *anything else*, they would be exempt from responsibility. Between these two extremes, agents have to deliberate between the few reasonable alternatives open to them; they have scope for action. The greatness of some inventors does not stem from the fact that they have produced innovations from scratch quasi *ex nihilo*. Rather, their admirableness is grounded in the fact that they produced these innovations against particular contextual resistances similar to the ones just mentioned. Responsibility means the fostering and training of dispositions and capacities to find morally acceptable pathways *within* these structures. The central focus on the people who initiate change within complex social and technological environments constitutes a humanist ethics of innovation.

3. Responsibility and Visioneering— Opening Pandora's Box

3.1 Up-Front Visions

In the previous chapter, the concept of innovation was introduced and critically discussed. It was argued that innovations are ambivalent and that in order to plan and shape innovation processes normative questions about the value of concrete novelties have to be answered and their realization has to be assigned to agents. Such activities presuppose the existence of agents who can form intentions through deliberations and shape emerging technologies that are embedded in complex socio-technical systems. Growing attention is directed at the way visions play a role in innovation processes long before concrete technological possibilities are realized or even shown to be feasible. By discussing the case of "visioneering" in the present chapter, I will first introduce a type of agent that attracts increasing attention in STS and TA research for supposedly playing a major role in innovation processes (Sand and Schneider 2017). Second, I will outline some of the preconditions for being a responsible agent as much as addressing the perils of attributing a certain type of responsibility.

The way visions came to the fore of TA and STS in the last decade is in itself worth an extended investigation.[18] Today, there can be no doubt about their significance for inter- and transdisciplinary research on technology. In recent years, the number of publications that dealt with the role of techno-futuristic narratives in the process of technological development has increased exponentially. Many of these publications reasonably point out that the role of such narratives for the course of technologies is crucial, especially in the field of new and emerging technologies (Sturken et al. 2004; Ferrari et al. 2012; Grunwald 2012c; Urry 2016; Grunwald 2017). At the early stage of a technological development (before any

---

18   An overview is presented in (Sand 2016).

© Springer Fachmedien Wiesbaden GmbH, part of Springer Nature 2018
M. Sand, *Futures, Visions, and Responsibility*, Technikzukünfte, Wissenschaft und Gesellschaft / Futures of Technology, Science and Society, https://doi.org/10.1007/978-3-658-22684-8_3

concrete artifacts appear), technological enterprises are shaped by agendas, visions, scenarios and other sorts of narratives that attract the attention of particular societal groups like scientists or the public in general, either in a positive or negative way (Borup et al. 2006; Grunwald 2009; Simakova and Coenen 2013; Grunwald 2017).[19] Marita Sturken and Douglas Thomas underline the ambivalent structure of the effects of technological visions which is their tremendous potential to raise both hopes and fears (Sturken and Thomas 2004, 2). Today, probably the most famous example for science at such an early stage is synthetic biology (Grunwald 2012b, 2015).[20] These upcoming techno-visionary sciences pose specific challenges (Grunwald 2013). Already at the very beginning of a technological innovation, policy makers and stakeholders call for orientation. Policy makers long for an assessment of technologies to prevent putting society at risk. Entrepreneurs, to give another example, are also interested in innovation processes. They are, inter alia, afraid to miss economically important developments. Assessing the techno-visionary sciences based on quantitative models is, however, nearly impossible since there is little knowledge about the actual technologies at this stage. Hence, it has been suggested that a reconstructive and hermeneutic understanding of the techno-visionary sciences and their impact could function as a substitute for the lack of knowledge and the many problems of predicting the futures of emerging technologies. Hermeneutic TA and other approaches aim to shed light on the early stage of technological development by investigating the role and contribution of visionary agents, by uncovering their motivations, by describing the social networks of techno-visionary enterprises and by analyzing the contents and the biographies of socio-technical visions and imaginaries (Grunwald 2009, 2014a).

---

19   This does not mean that visions cannot have an influence during other (later) stages of technological development. However, the focuses of the mentioned research lay on the emerging technologies at their very beginning. Visions are likely to play a crucial, communicate role in all stages of innovation processes. John Urry speaks, therefore, about the performativity of futures: "Visions of futures, whether dystopic or utopic, may indeed engender futures, as they are part performative and not merely analytic or 'representational'." (Urry 2016, 53)

20   Other examples of fields of research in which rather far-fetched visions predominate are discussed in (Sand and Jongsma 2017, 2016a).

Through these studies no less than the very "nature of these visions must be made transparent in terms of epistemic, normative and strategic issues." (Grunwald 2013, 23)

It has been suggested that the anticipatory study of the ethical issues that the envisioned futures might generate is an important aspect of their assessment. In general, these approaches aim to establish a critical way of dealing with socio-technical visions to uncover shortcomings of these narratives, to find alternative technological (or non-technological) pathways and develop an enhanced frame-work to govern the techno-visionary sciences successfully. Such research brings the significance of the technological narratives for the course of technological de-velopment to our attention and points out the necessity to develop an assessment regime for an adequate dealing with extreme uncertainty. In his paper *The Herme-neutic Side of Responsible Research and Innovation* Armin Grunwald summa-rizes:

> The factual importance and power of futuristic visions in the governance of sci-ence and in public debates are a strong argument in favor of the necessity of providing early publix and policy advice in the NEST [new and emerging tech-nology] fields [...]. Policy makers and society should knowmore about these pos-itive or negative visions and their background. They should understand what is going on scientifically and technologically, what is or might be at stake for future developments, where the grand challenges to society are in relation to NEST fields under consideration, and who might be affected by societal develop-ments based on NEST progress. In summary, this needs uncovering which mean-ing, values, and interests are hidden in the techno-futures being communicated. Thus, gaining a comprehensive understanding of the meaning of the NEST de-velopments under consideration forms the necessary basis for reflecting on re-sponsibility and is part of the RRI process. (Grunwald 2014b, 279)

Building on the assumption that visions have an effect on the course of the techno-visionary sciences, some authors have highlighted the normative dimension of this influence and their originators. It has been suggested and passionately outlined in several publications that visionary agents carry a special form of responsibility (Coenen 2011; Simakova and Coenen 2013; Cabrera Trujillo, Laura Yenisa 2014; Ferrari and Marin 2014). Visionary agents who present and promote socio-tech-nical visions increasingly attract the attention of scholars. Their activities are an interesting subject for STS, TA and practical philosophy. "Visioneering," as it has been called recently, is the umbrella term giving a name to those activities (McCray 2013). The belief in the responsibility of visioneers is getting increasingly prominent. At the same time, the presumed concept of responsibility is unclear or flawed. Therefore, a more nuanced understanding of responsibility will be developed in this article. Especially two questions will be addressed: Is causation a sufficient condition for the responsibility of visioneers and can they be held accountable for the consequences of their actions in a complex social system? In the following section, we will have a closer look at approaches to responsibility and visioneering. In the conclusions of this chapter, some pathways towards a better understanding of responsibility for visioneering activities will be sketched. The following considerations can also function as a methodic pillar to establish appropriate categories of responsibility for the social sciences (Arnaldi et al. 2014). Visions are up front in the technoscientific discourse. We should catch up here!

3.2   Wagging the Finger at Visioneering

In the last decades, determining the responsibility of social actors according to their contribution to a technological development was certainly one of the key motives for the extended study of responsibility. The increased complexity of human actions and their possible impacts on a global scale have given rise to the term "responsibility" to become an ethical key concept in the second half of the twentieth century (Bayertz 1995; Lenk and Maring 2001). Responsibility can be understood as the conceptual answer to problems of attribution for an increasingly unmanageable, globally acting mankind (Bayertz 1995, 4). If visionary agents play—as it has been suggested by many—a causal role for technological development, they are the right subject to wag one's finger at.

People who distribute visions, creatively construct new ones and present them to the world affect certain processes, like, for instance, changes in their recipients' beliefs. By performing such tasks, these people act and the consequences of their actions are events. Mass panics, legal resolutions, closing of businesses are events that might follow from a far-fetched vision presented with confidence by a scientific authority. Some authors assume that influencing technological developments is a sufficient reason to carry responsibility. Such a notion opens Pandora's Box of complex philosophical questions about responsibility. Before we open the box, we should first get a clearer picture of the most recent reasonings. Laura Cabrera has unambiguously ascribed responsibility to "visioneering" practices. She argues with clarity and passion. Hence, it is worth to have a closer look at the following quote of her *Visioneering and the Role of Active Engagement and Assessment*:

> Being a visioneer—regardless of which type we might mean—entails a high degree of social responsibility, as visioneers play an important role in social change, first and foremost because their visions are likely to give rise not only to enthusiasm but also to anxiety, and may well lead to the radical transformation of existing social arrangements, values and traditional structures. Furthermore, visioneers have an extended social responsibility, inasmuch as "visions have influence on the sciences' agenda", influencing—directly or indirectly—societal attitudes, perceptions and funding policies, "irrespective of their degree of plausibility, feasibility and speculativity" [...]. If we take this seriously, we can then see, as members of society, that we have a social responsibility to actively engage in visioneering as well as to question and challenge other visioneering projects. Furthermore, if we want visions to be sustainable and ethical, visioneering will also need governance, management and monitoring [...]. (Cabrera Trujillo, Laura Yenisa 2014, 205)

Here, Laura Cabrera clearly demonstrates why she thinks that visioneers bear a "high degree of social responsibility." She argues that visioneers play an important role for social change and recaps the potential of visions to raise anxiety and hopes. This—she continues—may lead to further societal transformations like the rearrangement of values, expectations and institutions. She specifically addresses the influence of visions on research agendas and research policies. It can be suggested that because of her phrasing "transformation of social arrangements," she also means that broader processes, like, for instance, the transformation or closing of a full disciplinary research area, can be affected. Eventually she deduces an imperative to contest reigning visions and engage in visioneering practices. In her line of argument she adopts the concept "visioneer" from the historian Patrick McCray (McCray 2012, 2013). McCray uses the term to characterize the various activities performed by people like the physicists Eric Drexler and Gerard O'Neill. Both pursue their far-fetched visions of space colonialization and nanotechnology more or less successfully. Their visioneering activities entail the public promotion of their ideas, raising awareness for their subjects, fostering a network of interested scholars and engineers and attracting the attention of potential donors and interested policy makers. In general, visioneering, as McCray understands it, also refers to the striving for the realization of the vision in the near future. Regarding the professional background of many visioneers, striving to realize a vision means conducting research to make progress. In this respect, visioneering is completely absorbed by ongoing scientific enterprises in various fields. The social status of visioneers is influenced by their academic background and their ranking as "experts" in their particular technological field (Sand and Schneider 2017, 24). It can be suggested that this social status affects the specific weight of their actions and presentations. Many people will likely consider them as authorities in questions about technology. Visioneers were very likely socialized in the scientific community and are of good standing there. McCray summarizes his understanding of "visioneering" as follows:

> To sum: visioneering means developing a broad and comprehensive vision for how the future might be radically changed by technology, doing research and engineering to advance this vision, and promoting one's ideas to the public and policy makers in the hopes of generating attention and perhaps even realization. (McCray 2013, 13)

McCray has given a huge diversity of actions a name: visioneering. Such a conceptual category helps to arrange and structure findings and evidence to pursue "telling a story." McCray's neologism is a helpful tool to systematically approach the variety of activities as, for instance, Eric Drexler, an "enactor" of nanotechnology, has performed (Rip 2006, 361). According to its structuring and heuristic function, the concept "visioneer" can be considered as an ideal type, as Max Weber puts it (Weber 1988, 190). Patrick McCray does not evaluate the activities he subsumes as visioneering. Whether visionary activities and the visionary engagement in technoscientific processes were performed responsibly is left open in his approach. In contrast, Laura Cabrera enters the normative level after having described the role of visionary practices in contemporary technosciences. In her reasoning, she seems to equal causal contribution to a certain process and moral responsibility. According to Cabrera, agents carry responsibility for those events they bring about, whether it be social transformations, a fundamental shift of the general atmosphere of the public sphere or the success and failure of research agendas. In short: at the fore of Cabrera's normative evaluation are the actions and effects of a specific type of social agent. Other authors who have contributed to this debate have pursuit a different approach.

In their article *Responsibility and Visions in the New and Emerging Technologies* Arianna Ferrari and Francesca Marin first point out that the massive influence of visions was until now mainly ignored (Ferrari and Marin 2014, 21). Ferrari and Marin state that such debates were largely based on the predictable application of a small amount of normative concepts (equality, justice, freedom, etc.) on emerging technologies. According to the authors, these few moral arguments left the interested party behind with dissatisfaction. The reasons were the underdetermined ideas of the emerging technologies in question (for instance nano- or enhancement technologies) and the epistemic problems of having a meaningful discussion about their future shape.[21] They propose that "technological visions require a different framework, which starts with the acknowledgement of their specific nature, i.e. as expressions of values and ideas of their promoters [...]." (Ferrari and Marin 2014, 26) Through discussing the etymology of responsibility, the authors point out that responsibility also means being aware of a certain commitment. Such a commitment is exemplified by the awareness of robust standards for empirical research, which is, according to the authors, sadly missed, for instance, in the debate on neuroenhancement. However, assessing the consequences of actions and considering them also plays a crucial role in their line of reasoning. When they discuss the possible effects of neuroenhancement, they suggest that visions might encourage our desire for improving our abilities and modify our relationships towards disabled people. Furthermore, they state that "the creation of these [enhancement] visions leads [...] to certain research programs being set up, which calls for justification of the allocation of these resources [...]." (Ferrari and Marin 2014, 30) Their conclusive statement follows the same argumentative scheme as Laura Cabrera's. The strong influence of visions on various social processes demands responsibility:

---

21   In order to be fair the authors could have mentioned that not only the "ideas" of those technologies are underdetermined. In many cases, also the ethical concepts applied are extremely vague.

> [...] these visions drive our actions, modify our roles (as parents, as citizens, as professionals, as policy-makers etc.), and influence research programs, scientific agendas and our understanding of our moral duties. By shaping our expectations, technological visions also influence our perception of the ethical issues at stake [...]. (Ferrari and Marin 2014, 33)

Elena Simakova and Christopher Coenen go down a similar road in their readable book contribution *Visions, Hype, and Expectations: A Place for Responsibility* (Simakova and Coenen 2013). They first consider some stages of the history of the nano-visionary discourse. Their short history tells of the birth of the nanotechnological vision and its "enactors" as well as its development into today's umbrella term "nanotechnology" (ibid., p. 244). The authors state that the current debate covers a multiplicity of coexisting (and sometimes incompatible) visions of nanotechnology. They put their finger mainly on the role of "nano" as a label. Through the strategic usage of that label enactors managed to homogenize different institutions (universities and entrepreneurs) to draw contours on collaborations between them and invoke the relevance of new research paths (Simakova and Coenen 2013, 259).[22] With a crushing statement, they summarize their argument as follows:

> [...] we put emphasis on the **role of "enactors"** [original emphasis], or key figures and organizations leading major technoscientific initiatives as a central strategy. Historically [...] discourses have ignored the variety of players involved in the production of societal outcomes and the relations between players. Apparently, such impoverished visions of technological changes, as opposed to evidence from empirical research on science and policy in practice, have been an incredibly instrumental approach in enacting and gaining support for large-scale technoscientific initiatives. (Simakova and Coenen 2013, 260)

The clear emphasis on the role of enactors—or in other words: the visioneers—to enact and gain support for technoscientific initiatives is one of the central elements in their conclusion. In a different article of Christopher Coenen, in which he deals in more detail with the transhumanist movement—an extreme techno-visionary

---

22   The homogenizing function of guiding visions was already pointed out in the 1990s (Dierkes et al. 1992, 42).

"movement" which seeks to overcome aging, dying and desires the radical enhancement of the human body—he also denunciated their *instrumentalization* of visions for reactionary purposes and marketing (Coenen 2011, 253; Nordmann 2013b). Coenen criticizes the usage of visions to distract from more urgent societal problems and to enchant the public with a fuzzy mix of scientific and fictional narratives entailing the possible result of decomposing science (ibid., p. 254). According to Coenen, the rationality of the scientific communication is at stake due to the transhumanists' *irresponsible* dealing with techno-visionary futures. In both articles the understanding of the origins and the development of visions such as transhumanism, a reflective assessment and means to provide alternative futures are presented as the cornerstones for a responsible dealing with the techno-visionary sciences (Simakova and Coenen 2013, 260).

The authors pointed their fingers at visioneering, focusing on the wide variety of actual and possible outcomes of these practices, which range allegedly from the shaping of our expectations to the destruction of the scientific communication or damaging scientific authority. Visioneers might as well distract from more urgent societal problems, enact new technoscientific enterprises and research branches, reshape the scientific landscape, homogenize collaborations, shape societal norms and values, and attract public attention in both positive and negative ways. In the following section, I will argue that even if a causal connection between the events just listed and the actions of individual agents can be detected, the conclusion on their responsibility should not be drawn.

3.3 The Compartments of Pandora's Box

The presented understanding of responsibility as a causal relation of a societal agent to certain events is central in current approaches to the responsibility of visioneers.[23] Let us put this notion in a more formal circumscription and investigate

---

23 Bringing a certain future about is one of the main purposes of visioneering, as Alfred Nordmann rightly points out (Nordmann 2013b, 89–90). But visioneering might also fulfill other purposes like motivating potential collaborators, staff members or becoming effective in public.

its philosophical underpinning. The claim is basically: if a person P has brought about an event E, he or she is responsible for E. The visioneer, who causes our values to change or sets up a research enterprise, fulfills the sufficient condition of this proposition. Hence—following a basic rule of rational argumentation—the conclusion can only be that he is responsible for our value change or the research enterprise. It is by no means clear, whether this is an appropriate understanding of moral responsibility. This section will shed some light on the concept of responsibility. Mainly four issues will be addressed that were either neglected or inappropriately dealt with in the previously discussed attempts on the responsibility of visioneers. It will be argued that the role of intentions matters for the ascription of responsibility. Other authors—as we have seen—have not mentioned intentions as a possible condition for responsibility, even though intentions play a significant role for responsibility. Related to this issue is the role that alternatives for actions play for responsibility. These two dimensions of responsibility will be addressed later on. They are also closely intertwined with issues of recklessness and negligence, which will be discussed thereafter. In a later section, the problem of accountability will be explored in more detail. Little attention was drawn on the question of how broader social developments can be traced back to particular agents. Since visioneering is just a part of more complex social processes, this is a huge issue. It will be argued that there is currently no reason that would justify the specific weight that is given to the roles of visioneers for that process.

3.4   Autonomy

There is an immense and technically complex philosophical discussion about the preconditions of being responsible. Persons—the main addressees of responsibility ascriptions—differ in several respects from other subject or objects that can be "causally potent." Animals, persons and hurricanes have in common that they can cause various things to happen. An animal can attack a zookeeper and a hurricane can destroy the coastline. In each of these cases, we customarily confuse a rather descriptive and a normative way of describing the happenings. We say, for instance, that the wildcat is responsible for the death of the zookeeper. Thereby, we do not mean that she is blameworthy (Pettit 2007, 173; Wolf 2013, 332). When we say in our ordinary language that something was responsible for an event, we conflate responsibility with causation (Wolf 1990, 40; Keil 2000, 2). However, only persons can be responsible in a moral respect. They are blameworthy in a sense in which animals and hurricanes are not and this is the original meaning of being morally responsible. It is, however, far from clear where the difference between these entities regarding their moral culpability stems from. Let us call this the external question of responsibility: "[...] whether, and if so why, any of us are ever responsible for anything at all." (Wolf 2013, 330)[24] This external question asks for a justification of the blameworthiness of persons in general. The internal questions of responsibility are for instance: From which age are we fully responsible? Are visioneers social actors that can be held responsible? Does insanity undermine the responsibility of a person? These types of questions are certainly linked to each other. The traditional answer that was given to the question "Whether, and if so why, is anyone of us ever responsible?", which dissatisfied many philosophers, is that persons are at liberty to do what they want: They are free, so to speak. David Hume presented a minimal concept of freedom when he wrote:

> By liberty, then, we can only mean a power of acting or not acting, according to the determinations of the will; that is, if we choose to remain at rest, we may; if we choose to move, we also may. Now this hypothetical liberty is universally allowed to belong to every one who is not a prisoner and in chains. (Hume 1975, 95)

Hume's notion of liberty as being free from external compulsion or coercion was adopted by twentieth century philosopher Moritz Schlick (Schlick 1962c, 148).[25] Yet, both Hume's and Schlick's reaonings are problematic. While the condition of being free to act might be important, other conditions must be presumed for being

---

24   See also for this distinction (Waller 2011, 3).

25   Here and in the following, I will cite from the second edition of David Rynin's translation of Schlick's Fragen der Ethik, which was originally published in 1939 (Schlick 1962a).

held responsible. The condition of being free from external compulsion applies to animals as much as to hypnotized, drug addicted and mentally impaired humans (Campbell 1951, 447; Wolf 1990, 10). They are all free from external forces—they are clearly not in chains. These examples show that the freedom to act cannot be sufficient for being responsible.[26] The hypnotized person and the drug addict are free from external compulsion, but they lack the ability to control their own behavior (Bieri 2013, 99). It is unsound to consider someone responsible who lacks the capacities to reflect upon his own wishes or preferences and to decide on such a reflected basis, which act to perform. As Gary Watson clearly points out the weighing of reasons plus the responsiveness to them are crucial for moral responsibility: "The idea that moral responsibility is crucially connected to the capacity to respond to reasons is a natural one." (Watson 2004b, 289) But the meaning of the expression "responding to reasons" is the crux in this statement and also constitutes the centre of debate between libertarians and compatibilists. Almost all philosophers engaged in the debate about responsibility emphasis the role of reasons and our openness to such reasons expressed then in our actions (Höffe 1993; Nida-Rümelin 2011, 17). But whether and in which sense persons are the authors of their actions and control the outcome of deliberative processes of balancing reasons is the crucial concern. The difference between Hume's type of liberty and the type of control that is at stake here can be considered as the difference between freedom and autonomy. The meaning and existence of autonomy—the type of genuine control we are looking for—was and still is highly questioned (Mele 1999). Only few philosophers doubt that anyone is ever responsible. They believe that autonomy is impossible because such a type of genuine control can only be explained through forces external to the agent or with reference to luck. In both cases the agent loses the ultimate control over his decisions (Pereboom 2007; Waller

---

26　　To be fair: Hume's and Schlick's reasonings are much more elaborate then it can be presented here. Both developed further strategies to justify the ascription of responsibility. Hume approved our moral sentiments to be an appropriate acid test for being responsible and Schlick believed that responsibility ascriptions also fulfill the function of influencing people's behavior to the better. Moral responsibility thus becomes a matter of being the appropriate object of praise or punishment. Both philosophers avoided employing a notion of genuine freedom. I will discuss their respective types of compatibilism in chapters four and five.

2011). To come back to the visioneers, one can say that they are at least not lacking a *certain* type of control. This weaker notion of control, as a "kind of freedom worth wanting" is emphasized by most compatibilists who believe that such weaker type is sufficient for being morally responsible (Dennett 1990; Kane 1996, 15). Scientists, engineers and visioneers are usually like other persons able to suspend their initial preferences, to turn into a state of mental deliberation, balance reasons and act according to what they consider as the best choice (Tugendhat 2007, 60–61; Locke 2008, 142). This is true even if it does not apply to all of their decisions (maybe not even to the majority of their decisions) (Douglas 2003, 59). Visioneers do not lack the power to execute their actions like the drug addict and they do not lack the capacities to reason. Albeit the complexity of the discussion about the origins of the ability that has just been described—to suspend initial preferences and balance reasons—it seems that it is a necessary condition of being blameworthy. Those attributes are the demarcating features between persons and other causally potent objects (like animals). In the context of visioneering, these conditions of responsibility have not been addressed before. And, if the type of ultimate control over these abilities is undermined by the laws of nature or explanatory deficits for such a type of control, then this applies to visioneers as to any other person. If the quest for autonomy remains unsuccessful, visioneers would not be the only bereaved.

3.5   Intentions and Alternatives

The authors we have discussed in the previous section argue that agents are responsible for the consequences of their actions. From this line of reasoning, it follows, for instance, that if a vision influences the public opinion, the visioneer would be responsible for this event, which is an intuitive idea. If, to consider another example, he or she brings about the termination of a research area, he or she would be responsible for that. Let us explore this idea of responsibility for consequences a bit further with the help of an example that has often been discussed in that context: an episode of Shakespeare's *Hamlet*. The death of Hamlet's father, the former king of Denmark, is the starting point of the famous tragedy. In the course of events that unfold after the death of the king, Hamlet kills Polonius, a more or less innocent person and advisor of the Danish monarch. At the beginning of the story, Hamlet's father appears to him as a ghost and informs Hamlet about his homicide caused by Claudius, the king's brother and successor to the throne. Hamlet, who is at first doubtful of his uncle's guilt, is convinced after observing Claudius's reaction during a theater play in which a similar assassination as that of his father is reenacted: Claudius leaves upset. Being certain of Claudius's guilt of his father's death, he settles with a plan for revenge. Additionally, Hamlet's frustration about his mother's role in the course of events is ever increasing. Being faced with her lack of understanding for his behavior, he outrages. Assuming that Claudius is hiding behind the curtains, Hamlet stabs through the curtain with a dagger. It is, however, Polonius who has been hiding there to spy on Hamlet and his mother. Polonius has deceived Hamlet, but he had nothing to do with his father's death. Hence, Hamlet is shocked when he finds out that he actually killed the wrong person. Later on, it is Claudius who seeks revenge and thereby kills Hamlet's mother unintentionally with a glass of poisoned wine that was meant for Hamlet. Both incidents are closely tied to questions of responsibility. Hamlet and Claudius *bring about* the death of persons they originally did *not intend* to harm. While Hamlet and Claudius are certainly blameworthy for killing, this is also the foundation of the story as a tragedy that their feuds both hit their beloved relatives hardest and unintentionally.

The examination of these cases shifts the focus from the externally related events of actions and their consequences to the *inner life* of the agents and their intentions. In each case, we must acknowledge that while they missed to harm their primary target, it is important that they both intended to kill *someone*, in Hamlet's case the King that he assumed to be standing behind the arras. Thus, one might say that there is a true description of the matters of fact that says what Hamlet did has been done intentionally and this description is: he intentionally attempted to murder Claudius, King of Denmark (and thereby killed Polonius) (Davidson 1980b, 46–47). One could argue that this sort of "distracted" intentionality in Hamlet's case is paradigmatic for the difference between a killing and murder (French 1984, 133). A murder is always the killing of someone specific, but a killing is not necessarily directed at the death of that person. Thus, soldiers might be considered as killers but not as murderers. Hamlet's reaction on his killing of Polonius lacks any apparent remorse, which makes it likely that he does not regret having killed Polonius. This, in turn, can be seen as evidence that he was willing to murder Polonius, too, although the evidence is not ultimately compelling (French 1984, 133). It is often tough or impossible to determine with certainty whether the description of a series of events as intentional is the right description of these events. From an external viewpoint, intentionality is opaque and not accessible. General attorneys often have a difficult job in rationally reconstructing people's intentions based on their relations to victims or other states of affairs that could function as evidence to answer questions regarding their responsibility. Important at this point is that intentionality demarcates the line between pure accidents, unreflected behavior, and actions. People (or to be more precise; people's bodies) can be involved in causal chains that are disastrous and yet lack any responsibility: they stumble, fall down the stairs, hit the light switch and a blown fuse produces a fire. Aristotle considers such incidents as something that happens *to* the agent and not *through* the agent (1110a 2–3)(Urmson 1999, 43). In contrast to mere behavior, it is essential to actions that they can *fail* (Hartmann 1996, 76; Keil 2000, 139).[27] It is not meaningful to say that someone has failed to cough or to stumble. An action can fail because it is directed at something—an end—and one can fail bringing this end about. Hamlet failes to murder the King, but one cannot meaningfully say that someone has failed to stumble down the stairs. An accident is not directed at an end—an accident is not directed at anything and, therefore, cannot miss such end either. It should be also clear that the success or failure of an action is often tied to the progression of natural courses. If I intend to grow the biggest tree in town, the success of my action is not yet enclosed when I have planted the seed (Janich 1981, 78). This is merely where my active engagement ends; then nature must take the right course: the coming winters must be mild and the storms rare and feeble. Here, luck can enter and redirect the course originally intended. This is set out in more detail in chapter five. Furthermore, in contrast to mere behavior one can meaningfully *request* someone to do an action (Hartmann 2005, 3). Because one cannot meaningfully request someone to stop coughing when having a cold, because one *cannot omit* coughing, or likewise sneezing, one is not responsible in such cases. The ability to do something else—

---

27 Searle puts this in nuce: "In general then, intentional states have 'conditions of satisfaction'." (Searle 1991, 60)

or, in other words—to do otherwise (which might in certain case be omitting to do something) is also central to the notion of action. Hamlet failed to murder the King. But he certainly is not exempt of responsibility. *Willing* to murder someone and, then, pursuing this end is clearly morally blameworthy, even when it misses its ultimate target. This is the right description of the situation and it is a description that sates clearly that Hamlet *did something intentionally*: He willed and tried to murder someone.

Less dramatic, but equally significant regarding the relation between causing and intending are the following cases: We can easily imagine daily routines or behavior that stand in a causal relation to disastrous events (in the near or far off future) which we did not intend because knowledge was *fragmentary* or *improperly reflected*. This opens the possibility that we do not bear responsibility at all for some things which we could not have known (early industrialists could not have known about climate change) and for others our responsibility does not concern the indirect effect of our actions but our acceptance of the entailed risks. Having acted out of negligence or recklessness can be the verdicts in these latter cases, as will be discuss in the next section. None of the authors that have been discussed in the previous section has spent a word on the agents' actual intentions although they seem to play such a fundamental role. We have to ask whether and how these considerations apply to visioneers. Visioneers perform various actions, for instance fostering networks, publishing books, etc., with a huge variety of impact, but their intentions for either of those actions are neglected and largely unknown. For philosopher John Mackie intentions built such an important condition for responsibility that he introduces his "straight rule of responsibility." (Mackie 1990, 208) Mackie states that we are responsible for all, and only, our intentional actions. The illuminative insights of Mackie's discussion—especially his accentuation of intentions—are certainly beneficial to gain an appropriate understanding of responsibility. In Mackie's straight rule of responsibility the lack of sufficiency of the causal criterion is emphasized. Not only causation is a necessary condition for attributing responsibility, also intentionality is required (Wolf 1990, 40; Douglas 2003, 61). Furthermore, Mackie provides an interesting account for the role of alternatives for holding someone responsible. He uses an example of Aristotle to highlight the value of having a choice. Thereby, Mackie reminds us that an action can appear in a completely different light if we know about the alternatives that were available to the agent. Such insight into the context of an action has an impact on our *moral evaluation* of the action. Mackie discusses the following case that originally introduced by Aristotle in his *Nicomachean Ethics* (1110a 5–15): Imagine a ship is overtaken by a storm. The captain of the ship has to decide whether his crew should jettison the cargo or leave it untouched and hence risk sinking and dying. If he is a rational being, he will certainly choose to get rid of the cargo. Back on land, his business partners will likely make him responsible for the loss of their goods. However, when he describes the situation and explains what made him decide to jettison the cargo instead of risking the lives of his crew members, the anger about his decision will ebb away. It is likely that the captain will even be honored with appreciation for his humane decision. The central message of this example is that an action, which is at first sight disputable, might become heroic if it was performed instead of the only, intolerable alternative. We could call this— following Gary Watson—the "avoidance opportunity"-condition of responsibility (Watson 2004d, 279; Braham and van Hees 2013, 607). Acting responsibly means choosing rationally between given alternatives.[28] Thus, responsibility as being "causally potent" and blameworthiness are clearly separated. The captain is causally responsible for the loss of cargo, but he is not blameworthy due to a lack of reasonable alternatives. Clearly, whether an option A is more reasonable than another option B is a normative question (see also section 5.6). I will explore the possibility of separating blameworthiness and being responsible (judging someone

---

28   A lack of reasonable choices is sometimes interpreted as a lack of freedom. Kant sometimes speaks convoluted about being determined by the will (Schönecker and Wood 2002, 177). This terminological shift should be handled carefully because here the notion of freedom differs again from the freedom to act and from autonomy, both of which were introduced in the previous paragraph. Rationality does not determine the choices of agents. It offers guidance like a roadmap without forcing an agent to perform a particular action. The captain in our example is not determined to throw the cargo overboard. Yet, he is "pushed" by the unacceptability of his alternatives. That rationality does not determine in the same manner in which the laws of nature would determine gets obvious when we consider that rationality does not disclose alternatives. One can act irrationally, but one could not act "against" the laws of nature, they undermine alternative possibilities (Keil 2007, 53).

to be responsible) in more detail when discussing retributivism, effect compatibilism and moral luck in later chapters (Smith 2007).

It should be obvious by now that balancing reasons and intentionality are closely connected. One could argue that people, who have balanced reasons, have set themselves ends to pursue. They have—in other words—formed intentions. If asked for those intentions they could (at least in principle) disclose them. Many authors writing about responsibility mention that the prefix "response" in the term "responsibility" refers to its Latin origins "respōnsum" which means to answer or to reply (Watson 2004a, 8; Nida-Rümelin 2011, 12; Grinbaum and Groves 2013, 120; Ferrari and Marin 2014, 28). People who are suggested to be responsible can usually respond to the question: "Why have you done this or that?" Agents can justify their actions *ex post*, if they have deliberately choosen to pursuit certain ends (Smith 2012, 578). Answering to questions about the reasons for actions is the most basic speech act included in the "responsibility"-game in which only competent humans can participate. By asking for reasons external observants who are on the verge of holding someone responsible try to understand the motivations behind someone's actions, extend their knowledge about the decisional situation from the perspective of the agent, and assess the quality of her reasons (Bieri 2015, 96). It might turn out that alternatives that appeared to be open from an external point of view were unacceptable for a particular agent given his biography or particular relation to other involved agents. The captain of our cargo ship appears rather cowardish, for example, when it turns out that he jettisoned the cargo not to rescue his crew but because he gambled about a large sum of money if he returned without cargo and that is an information that is prima facie inaccessible for external observants and judges. We have to ask him for this—without being able to guarantee that he will honestly answer. Gary Watson writes: "Holding others responsible takes the form of requiring them to answer for their behavior, to give an account of themselves in light of their apparent violation of certain expectations, and, furthermore, to respond appropriately in case that account is unsatisfactory [...]." (Watson 2004a, 8) "Answerability," as this condition has been called in the literature, cannot substitute the straight rule condition. On the contrary, answerability rides piggyback on the condition of deliberately choosing one's ends and intending to do certain things (ibid., pp. 8 f.). Only if an agent has gone through such a process he can answer to a demand for justification with something more meaningful than a drilled response.[29] This is not the case, as outlined above, when an agent accidently slips or when being asked for things beyond her control (like her body-height (Smith 2012, 578)). The disclosure of reasons for action puts external observants such as judges or other members of the moral community in a better position to evaluate someone's responsibility, draw conclusions about the blameworthiness of an action, and determine the possible *degree of sanction*. Furthermore, when actions are too trivial, we will not demand to answer an agent for anything and still the basic notion of *being* responsible for this trivial action is suitable.

The outlined ideas and Mackie's straight rule condition of responsibility can be applied to visioneering. Visioneers have a finite set of choices when they act. If they are entrepreneurs and run a company that does not prosper, there are only a limited number of means to react to it. Presenting a new vision to the staff is maybe not the worst solution to commit and motivate employees towards a new research pathway. In theories of entrepreneurship dedicating yourself and the staff to a vision has been considered as a reasonable and important tool for the success of a company (Byers et al. 2011, 54). There is little knowledge about the context of actions and the range of alternatives visioneers have for their innovative decisions. This is an important connecting point for social sciences. Responsibility can only be ascribed to them if there is knowledge about the alternatives that are open to them and if they are taken into account in evaluation. The social sciences can fill this blank. It can be assumed that visioneers are embedded into a force field of competing social (and corporate or more generally institutional) influences just like other actors in technoscientific fields. They have to maneuver along a narrow path to attract funds and build up new alliances.

Coming back to Mackie, there is at least one fly in the ointment. Mackie takes so much care to develop a comprehensive account of the conditions for responsible actions that he neglects the question why actions and only actions should stand in

---

29   Clearly, we sometimes rationalize activities that were performed thoughtlessly afterwards. Or we search for stronger reasons than those that originally persuaded us. These are strategies in order to defend oneself against moral appraisal and they lead to justifications for decisions that have not been deliberately made. If entertained whole-heartedly they can also deceive one's memories of how things really went.

the fore of responsibility ascriptions. Agents perform actions and it is plausible to hold them responsible for them. Hamlet *intended* to murder someone and in executing this plan he even killed someone for which he is responsible. But why should we narrow responsibility to actions? Attitudes or character traits in general are also proper candidates for which agents can be responsible and they should not be excluded beforehand. I will outline some examples in the following section that make such extension of our understanding of responsible agency more plausible and also how it can be incorporated in Mackie's approach. Furthermore, in my previous discussion there is an important dimension that has not been considered yet, but which is crucial to understand responsibility in technological processes. How can we actually prove that someone plays an important role for a technological process? Before I discuss this in more detail, we will consider an important aspect related to responsibility and intentionality: negligent and reckless behavior.

3.6   Recklessness and Negligence

Scientists, engineers and visioneers are usually like other persons able to suspend their initial preferences, to turn into a state of mental deliberation, balance reasons and act according to what they consider as the best choice (Tugendhat 2007, 60–61). This is true even if it does not apply to most of what people do as, for instance, how to get dressed in the morning (Douglas 2003, 59). One does not deliberate in the morning whether to put on the right or the left shoe first. Regarding the existence of a vast amunt of habitual behavior and unreflected routines one might argue that I have simplified the concept of responsibility by adopting Mackie's straight rule. One might suggest, in contrast to the straight rule as proposed, that there are clear instances of actions with terrible, unintended consequences for which an agent can be judged responsible and is usually held responsible. I have employed intentionality to distinguish accidents and mere behavior from action and argued that people are responsible only for those things they intentionally do—for their actions. Is there behavior that can properly be considered as action and for whose results one can be responsible? Consider, for instance, a person who drives too fast and runs over a pedestrian who dies. We will assume—for the sake of argument— that he would have been able to avoid hitting the pedestrian, had she driven as indicated by the traffic signs. This person—as we suggest—has *not intended* to hurt or even kill *someone* (in contrast to Hamlet). If intentionality were a necessary condition for being responsible, we would have to conclude that this driver is not responsible for the pedestrian's death because she did not intend this to happen. This is a conclusion that Jay Wallace seems to entertain when he writes: "[…] one can be said to have complied with a moral obligation only when there is present a relevant quality of choice. Someone who inadvertently bumps into me, thereby knocking me out of harm's way, has in no sense complied with the obligation of mutual aid […]." (Wallace 1994, 132) In reality, however, we do hold reckless drivers responsible (Simons 1999; Smith 2005, 263). Before the law, such cases of *negligent homicide* are not as severely punished as murder, which has been understood as the intentional killing of someone in the previous section. However, one also has to expect stiff penalty in cases of negligent homicide. In addition, our moral reactions following up on such incidents are severe: We will certainly blame and reprehend a person who caused a fatal accident due to speeding.

Do we have to give up this practice because these responses are inappropriate or do I have to drop (or at least substantially refine) the straight rule of responsibility because these examples indicate that there is a responsibility for *unintended* harm? Neither is the case. If we interpret negligent and reckless behavior properly, we can adhere to the intentionality condition and (at least sometimes) hold persons responsible who bring about harm unintentionally. The key to understand why someone can be responsible for the death of a person without having intended this is to acknowledge that negligence involves a kind of *indirect intentionality.* In order to understand the behavior considered as negligent, one has to take into account the history that *precedes* the speeding of the negligent driver. In the course of getting a driver's license one is repeatedly informed about the dangers and responsibilities of being a road user. One is taught how to interpret street signs and how to consider wheather conditions when driving. One is also sufficiently informed about the results of misbehavior on roads and about fatal accidents statistics. This means that during the time of getting a driver's license one must *reflect* on the newly gained "power" and increased responsibility that is accompanied with the right to drive a car and on the possible effects of driving too fast. The quality of this reflection and gained competence is being tested before a driver's license is issued. Therefore, by driving too fast, the negligent driver—being fully aware of his substantial responsibility—deliberately and *intentionally ignored* the inrcreased *risk* of harming other traffic participants. This is the intentional act preceding the negligent behavior for which she is rightly held responsible. When making her responsible we do so arguing that she could have known that the likelihood of such fatal occurrence increases vastly when speeding. Because of the presumable awareness of risk Mackie says that the fatal crash has been "obliquely intended." (Mackie 1990, 211)[30] With regard to the negligent driver it seems to be paradoxical at first that a person who negligently drives and hits someone is approached differently than a person who drives similarly negligently and does not run over anyone (Mackie 1990, 212; Nagel 1991b). This problem of moral luck will be discussed in more detail in chapter five.

A few remarks about this argument for being responsible when behaving reckless or negligent should be added. First, it is not easy to distinguish between reckless and negligent behavior involving *indirect* intentionality and behavior that does not entail *any kind* of intentionality, which is purely accidental so to speak. Consider, for instance, a person who turns on a light switch that blows a fuse and burns down the house as a result. This was neither intended nor as a likely result of the act negligently accepted: One could not have known it beforehand. But what about more ordinary behavior: is it negligent regarding one's own well-being to eat fast-food which is suggested to be unhealthy in the long run? Or, is it negligent regarding the environment to eat meat? Epistemic and normative question intertwine in the application of the concept "negligence" to particular behavior. As an approximation, one could first distinguish the conscious acceptance of a *high probability* of the occurrence of harm or the probability of *major harms* as recklessness to negligence, which can be understood as the conscious acceptance of *less probable* or *minor harms*. It seems elegant to connect the degree of responsibility to the severity of negligence in terms of probability of an event and the harm that might occur (which are both factors in a concept of risk) (Simons 1999, 57–

---

30   Ralf Stoecker is dissatisfied with Mackie's straight rule although he offers an interpration of its application to negligent driving (Stoecker 1997, 359). I think that his dissatisfaction stems from a misunderstanding of the problem of moral luck. This appraisal is supported when considering Stoecker's causal theory of action (and responsibility). His proposal does not solve the problem of moral luck, because it is beyond the driver's control (causal power) to avoid hitting the pedestrian.

61). Therefore, a drunken car driver is negligent but a drunken pilot even more so because of the multitude of harm and the increased likelihood of its occurence he can cause. Similarily, a person who drives too fast on a solitary field is less blameworthy because of the lower probability of harming someone than someone who drives too fast in a crowded pedestrian zone which would be grossly negligent or according to the previous reasoning: grossly reckless. How likely is it then that one harms future people by eating meat and how severly is the harm future generations will have to sustain? Moreover, even if there were more robust knowledge about those aspects, would it suffice to consider "meat eating" as negligent behavior? In a world that is increasingly complex even the most unsuspicious behavior entails a certain degree of risk (Urry 2016, 63). I will not provide a solution to these difficult problems. For the present purpose, it is most important to emphasize that there are clear cases in which the application of the concepts recklessness and negligence is reasonable. Because people can deliberately and volitionally choose to behave *carefully*, *attentive*, and *risk averse* a sort of indirect intentionality underlies or precedes much of people's behavior (Pauen 2004, 101). This is clearly the case when someone drives too fast. Therefore, the straight rule of responsibility as outlined before certainly applies to such cases.

Second, raising the awareness for the possibility of indirect intentionality leads us to a more nuanced picture of the wide scope of things for which people can be responsible. This includes not only actions as deliberately and consciously performed tasks but also *attitudes* and *character traits*, which can be volitionally transformed. Consider, for example, people's attitudes towards knowledge: some people believe they already know everything and cannot be enlightened anymore. Such people are *ignorant* about their possible lack of information and about possible sources to extend and refine their knowledge and reasoning (van de Poel 2015a, 22). If it would be publicly discussed and substantiated that meat eating has likely and terrible consequences for future generations an ignorant person might self-righteously continue eating meat without having a bad conscience. Afterwards, such person might try to defend this practice arguing that she did not know meat eating is harmful and that she never intended to make anyone suffer. This, however, is not an excuse she can utilize: An *ignorant* person is clearly blameworthy in such case because it is a result of her preceding ignorance that she lacked the common information about the consequences of meat eating. This ignorance is an attitude she *consciously employed* and could have dropped. Angela Smith provides a similarily pertinent example in which the intentionality-condition of the straight rule of responsibility and the responsibility for character traits are problematized at the beginning of her article *Responsibiliy for Attitudes: Activity and Passivity in Mental Life*:

> I forgot a friend's birthday last year. A few days after the fact, I realized that this important date had come and gone without my so sending a card or giving her a call. I was mortified. What kind of a friend could forget such a thing? Within minutes I was on the phone to her, acknowledging my fault and offering my apologies. But what, exactly, was the nature of my fault in this case? After all, I did not consciously choose to forget this special day or deliberately decide to ignore it. I did not intend to hurt my friend's feelings or even foresee that my conduct would have this effect. I just forgot. And yet, despite the apparent involuntariness of this failure, there was no doubt in either of our minds that I was, indeed, responsible for it. (Smith 2005, 236)

Although Smith provides a different argument, she arrives at the same conclusion than I do: Forgetting a birthday is clearly a failure attributable to a person for which she can be held responsible. My own approach that refers to Mackie's straight-rule introduced above is what Smith calls the "prior choice" view for which it is "essential for attributions of responsibility [...] that we be able to trace the development of an attitude to a person's own prior choices or decisions." (Smith 2005, 239)[31] Her friend's sadness about forgetting her birthday is best understood as a result of her *forgetfulness* which can be considered a character trait. When someone realizes that she is forgetful, we expect her to write down reminders about upcoming birthdays, to buy a calender or use other means to deal with this *vice*.

---

31   Smith dismisses the "prior choice" view. In contrast, she proposes a so-called "rational relations view" according to which a person is responsible for things (attitudes, actions, emotions) when these things "reflect her evaluative judgments." (Smith 2012, 578) Although this account also properly captures negligent behavior as something for which agents are responsible, the definition of responsibility offered is much too broad. According to this view, I might also be answerable for the wrongs of other people who share certain aspects of my (political) ideology that reflect my evaluative judgments.

While it might not be within an agent's control whether one is born as a forgetful person (which is what Nagel calls a matter of constitutional bad fortune (Nagel 1991b, 33)), it is a matter of *deliberate choice* whether to accept this trait, develop strategies to lower its impact, or try to overcome it eventually (Smith 2005, 266). Therefore, negligent behavior can be blameworthy and does not provide a counterexample that devalidates the straight rule. By emphasizing forms of *indirect intentionality,* we advance our perspective on responsible agency. It seems that it is in itself a demand for responsible personhood to develop strategies to deal and overcome character traits such as forgetfulness. This is what might be termed the *aretaic* dimension of responsibility (Watson 2004d, 266; Smith 2012, 576). The discussion of negligent behavior as indirectly intentional provides a smooth bridge from a focus of harmful or benevolent acts to the responsibility for attitudes and character traits as indirect intentional. Both can be deliberately shaped and ameliorated by an agent. These cases push one to a more advanced picture of negligent behavior taking *preceding decisions* of an agent into account. Before ending this chapter, we should dwell a bit further on the causal condition of responsibility and its application to visioneers.

3.7 Accountability[32]

There is a common theme in all approaches on the responsibility of visionary agents, namely that visioneers *contribute* to technological development. This "paradigm" seems to be widely accepted. Examples of influences based on visionary practices have been presented: Eric Drexler for instance has been considered as the "Apostle of Nanotechnology." (Amato 1991) He is a preeminent representative

---

32   Unlike other authors I understand accountability in the following as the tracing of actions as the origins of certain events. The proper German translation of accountability in my understanding of the term is "Zurechenbarkeit" (not "Zurechnungsfähigkeit"). If someone should be held accountable this connection is necessary, but blameworthiness requires a rather deeper understanding of the person as author of her acts. In the literature the term "attributability" is sometimes preferred (Watson 2004d, 272). Grinbaum and Groves speak of "imputability." (Grinbaum and Groves 2013, 120)

of the visionary origins of nanotechnology (Rip and Voß 2013; Grunwald 2014b, 2017, 6). Without denying that Drexler was an antecessor and idea giver for this technology one could ask: What did he actually effect? Did Drexler cause the launch of the National Nanotechnology Initiative (NNI) in the early two thousands? Is he responsible for the American mega project to emerge? Or would he be responsible if eventually the horrific "grey goo" would swallow the earth? These questions are purposefully phrased rhetorically because they presume a very simplistic picture of technological processes. Technological systems can be described as highly complex. Various stakeholders, research enterprises, companies, politicians and consumers contribute to the actual shape of a technology. STS have shown that technologies are socially constructed and that we can generate complex narratives of their development.[33] This feature makes it hard, if not impossible, to account a particular person for an abstract and complex development like a whole research area or the norms of a society. Accountability is not always such a big deal for responsibility. In the previously considered example of Hamlet for instance, we are not facing such "epistemic blind spot." In fact, the story is told in such a way that the problem of accountability does not arise. It entails one agent and one clear effect of his particular action. Accounting Eric Drexler for a particular development, however, reveals such an "epistemic blind spot." There is little knowledge about the influence of agents on broad social processes like the change of certain norms or the development of a new research area. The claim of some contemporary analysts of visioneering responsibilities was that visioneers stand in a causal relation to such wider social developments. Such a claim is, however, hard to defend. To exemplify this point we can discuss nanotechnology. Besides Drexler, numerous other people and organizations were involved in the development of nanotechnology (Paschen et al. 2004; Nordmann 2006). To name just a

---

33   This process can be described as a system entailing innovation as an emerging product of the process. To understand responsibility it is important to note that such a systemic notion of innovation processes neglects individual contributions to this emergence. Performativity as the source of innovation processes contrasts with the interpretation of those processes common in some economic theories as endogenously driven (see chapter two). This conflict of "emerged" and "produced" effects reappears in theories of collective actions and their outcomes.

few: Richard Feynman, Bill Clinton, Mihail Roco and William Bainbridge, etc. What has been their particular contribution to the current reality of nanotechnology? They certainly stand in a relation to its current shape, but this notion is excessively imprecise. The construction of visions and their succeeding technologies is as much as the construction of technologies in general a collective enterprise and many players contribute to it (Bijker et al. 1987; Swierstra and Jelsma 2006). Hans Lenk has clearly pointed out the problem we are discussing:

> With regard to moral assessments and judgments of responsibility it is difficult, if not impossible, to attribute in fields of synergetic and cumulative effects, in view of the mentioned systems effects, the responsibility for the application and implementation of detailed aspects of progress to just one individual technologist or researcher. If the development and acceleration do depend on a multiplicity of mutually escalating interconnections and interplays, it is not possible to attribute the responsibility to just one person. (Lenk 2007, 46)

Visions get adopted, deformed and rewritten. Their particular impact may be small, but cumulated with other causes wider social processes may be brought about. Since so many social agents contribute to this process, following the crude logic of causal sufficiency, *all of them* are responsible. A special responsibility of the visioneer cannot be derived from the general assumption that they are relevant in the development of new technologies. This problem does not imply the impossibility of responsibility in general. It rather provides a reason to think more extensively and clearly about responsibility and causal chains in complex social developments (Lenk 2007, 46). Either an extended knowledge on the factual importance of visioneers for the technological development or a more elaborate concept of collective responsibility is needed (Grunwald 2000b). Thus far, such knowledge is largely lacking because research on the social phenomenon of visioneering is underdeveloped. The assumption that they are important agents for the development of emerging technologies—as, for instance, Laura Cabrera assumed—is currently indefensible and so is the conclusion on their special responsibility.

3.8   Future Challenges in Dealing with Futuristic Narratives

Many authors have provided valuable contributions to the understanding of visioneering practices and their role for technological developments in recent years. It has become obvious that visions of future technologies are a factor in shaping research agendas and their public acceptance. It also has to be acknowledged that technological innovations are visions first before coming into material existence. It is therefore an important step forward to discuss the responsibility of agents that provide visions of future technologies (Grunwald 2017, 9). In order to progress this line of reasoning, the aim of this essay has been mainly to criticize present considerations of visioneering responsibility. Through analyzing the current approaches of visioneering and responsibility, an advanced understanding of responsibility was developed. It has been argued that a certain type of control is needed to be held responsible. Usually, no one is denying that people who are active in the development of new technologies have such capacities. However, the brief overview presented in this paper has uncovered how closely the external and internal questions of responsibility are intertwined. The biggest and trickiest compartment of responsibility—an appropriate understanding of the role of autonomy—is often neglected in normative discussions in STS and TA. Furthermore, it has been argued that the causal relation to an event might be a necessary, but not a sufficient condition for being responsible (Bayertz 1995, 14). The discussion of some examples in this paper has revealed that intentions and alternatives play a crucial role for being held responsible. Solely the causation of certain events is not sufficient to be held responsible. As Heather Douglas writes: "[…] we are not held morally responsible for all the things in which we play a causal role." (Douglas 2003, 61) Visioneering is defined as a set of different activities and it can be questioned whether all of them are planned in a forward-looking manner. Hamlet's killing of Polonius illustrates the possibility of accidents even when practices are permanently reflected. This theme will reoccur in my discussion of the problem of moral luck in chapter five.

Therefore, in a theory of responsibility, both aspects—causing and intending—must be included and given adequate weight. Furthermore, it was argued that alternatives to visioneering also play an important role. We have to find out whether there are at all possible alternatives to providing visions to maintain research branches and technological pathways. An important step to realize the weight of this argument can be made by having a look at the many promises that are made by STS and TA when research proposals are written, as research is largely funded because stakeholders and policy makers expect beneficial outcomes. If such promises are the only way to maintain funding, it is reasonable to pledge. This case is not substantially different to other visionary practices. TA's and STS's goal of finding methods to establish a more democratic way of doing science is no less a promise than developing a groundbreaking new technology. The third compartment that has been discussed in this section dealt with the issue of accountability. Technologies are socially constructed. Visioneers are not the only factor in the development of emerging technologies and it can be doubted that they are as important as some researchers believe. When we consider very broad social processes such as the change of societal norms, it is nearly impossible to trace those developments back to one single origin (van de Poel 2015b, 51).[34] Especially the influence of visioneering practices on such processes is largely unknown.

Yet, the conclusion that responsibility fully vanishes under the opaqueness of causal complexity is wrong. It is rather an assignment for the social sciences to continue to study the emergence of visionary practices and their role for the technological development. This is the challenge for future dealings with futuristic narratives. We have to study the intentions of visioneers, their alternatives and the effects of their actions thoroughly to find out whether they are responsible. Surveys have to uncover which motives visioneers have to promote their visions in a particular way and not in another. Furthermore, we need a precise picture of the collective and corporative settings in which visioneering is performed. Visioneers work for universities and research facilities in various positions. Their influence on the agendas of these institutions depends on their position in the corporation.

---

34   The historical sciences are familiar with that problem. "Mono-causal" explanations of historical events have often been criticized. Notable is the attempt to explain the right-wing tendencies in Europe at the end of the 1920s as a result of the Great Depression. Such an explanation is a reduction of a complex social phenomenon to a singular cause (the crisis).

Many institutions incorporate responsibility according to their ladders of hierarchy. Higher positions often automatically entail a higher degree of responsibility according to an increasing power to influence overall agendas. While corporative responsibility does not necessarily concur with moral responsibility, it is, however, a form of responsibility that might provide a fundament for resentment or disciplinary measures according to certain corporative statutes, albeit with a different addressee (Maring 2001; Swierstra and Jelsma 2006). I will address these issues in detail in chapter six.

In my critique of Mackie's approach, I mentioned that his strong focus on actions as the only source for responsibility ascriptions is unjustified. Providing visions and using them as instruments to achieve certain goals can make people in innovation contexts responsible in the light of the previous discussion. Much more than performing responsible actions it might as well be the standard of being a virtuous person that innovators (and any other person) have to live up to. When we think of responsibility, we still tend to apply the previously considered model of *ex post* responsibility which is similar to our legal understanding of responsibility: When someone does the wrong thing and harms others he or she has to respond to accusations and possibly take punishment (Grinbaum and Groves 2013, 120). This model focuses on the wrongdoing of a person and not on his or her moral character. Visioneers perform certain actions that cause more or less predictable events. However, despite these particular activities and their respective outcomes, visioneers exhibit characters traits that are neither properly understood nor reducible to their particular actions. Promoting a vision convincingly over month or years and standing up for such an idea requires eagerness and devotion. Many innovators are persistent and creative in reflecting on how to realize their vision and technological change in general. They often stand behind their ideas in a charismatic fashion. These character traits can well be considered as *virtues*. To pursue these goals successfully, however, the innovator has to ignore skepticism and sometimes override external obstacles rigorously. Hence, innovators are (and maybe must be to become successful) somehow obstinate and stubborn about their possible impact on society. The character traits of innovators have escaped the attention of most of the previous moral assessments in innovation ethics. It is important to rectify this shortcoming and scrutinize the virtues and vices of visioneers in order to understand their responsibility. Furthermore, we must not assume that the responsibility for technological developments rests solely on the shoulders of some more or less eccentric individuals. There are effective governance mechanisms that support the shaping of innovation processes in a morally desirable way and democratic societies invite their citizens to participate in these processes. However, their transformation presumes agency, too (Sand 2016, 345). I will outline both aspects in more detail in chapter seven. When we think about behaving responsibly, we are in general well acquainted with the phenomenon of decision-making and acting. We have genuine inclinations towards certain ends like eating chocolate. We reflect upon the consequences of pursuing this end (gaining pounds) and then decide against eating chocolate. Nothing seems less questionable than our ability to make such and similar decisions. Yet, from a naturalist point of view, this ability appears to be embedded in the wider working mechanisms of the universe and the law like conjunction of events occurring in it. From such perspective, the whole idea of agency loses its intelligibility. Must we assume that visioneers cannot be held responsible for whom they are, if they cannot help being who they are—if their decisions are merely the product of preceding events as the naturalist picture suggests? How can we give a positive account of the ability to act? In the following chapters, I will address these and related questions.

4. Responsibility, Determinism, and Freedom

4.1 Introduction

The previous chapter explored the role of visioneers for innovation processes and their use of futuristic narratives in order to raise public attention for technological possibilities and foster networks of patrons. Recent arguments in favor of attributing responsibility to this type of agents were critically assessed. I have outlined some pitfalls regarding the attribution of responsibility for consequences in complex socio-technical systems. The causal contribution of visioneers to emerging socio-technical dynamics is far from clear as much as their foreknowledge about the likelihood of the disastrous consequences supposedly arising from their actions. However, another aspect has only insufficiently been addressed, and this issue does not only concern visioneering but also our understanding of agency in general. What if responsibility presupposes capacities that *natural* beings cannot possibly have? Since agency plays a central role in the present enquiry, these questions deserve to be discussed more extensively in the following chapters.

Do we have to have a genuine free will in order to be morally responsible? What does "genuine free will" mean? Does it make sense to punish or blame people, if they could not have acted otherwise? Moreover, what does acting otherwise actually mean? In this branch of philosophy—the freedom of will problem—the most intricate difficulties from the philosophy of mind, theory of action, and practical philosophy culminate. Given the maze of superb arguments that have been provided in this debate over the centuries, one can easily get lost and end up dissatisfied or even pessimistic about the possibility of ever finding a satisfactory solution. Thomas Nagel once expressed such pessimism when humbly admitting that he often changes his opinion about the problem and noticing that "nothing approaching the truth has yet been said." (Nagel 1986, 137) Neither will I attempt to solve all the issues accompanied with the freedom of will problem in the present text. Such an attempt is far beyond the scope of this work. Instead, I will focus on

© Springer Fachmedien Wiesbaden GmbH, part of Springer Nature 2018
M. Sand, *Futures, Visions, and Responsibility*, Technikzukünfte, Wissenschaft
und Gesellschaft / Futures of Technology, Science and Society,
https://doi.org/10.1007/978-3-658-22684-8_4

the problems that have close connections to normative issues, for instance the problem of moral luck, and retributive justice. In both cases the issues about freedom of will and compatibilism are, at first sight, closely intertwined with the notion of just desert. Thus, it seems that it cannot be fair to attribute moral responsibility (through praise and blame) for actions and their consequences, if these are beyond people's control, and, so the argument continues, if determinism were true, we were all devoid of this capacity and not morally responsible. Therefore, we cannot accept a compatibilist standpoint that assures us that to be an appropriate addressee of praise and blame, a much weaker notion of freedom is sufficient that applies when someone is the bearer of a motive or has been free from external compulsion.

Charles Campbell emphasizes the connection to normative interests that guides the quest for genuine freedom and argues that "the problem of free will gets its urgency for the ordinary educated man by reason of its close connection with the conception of moral responsibility. When we regard a man as morally responsible for an act, we regard him as a legitimate object of moral praise and blame in respect of it. But it seems plain that a man cannot be a legitimate object of moral praise or blame for an act unless in willing the act he is in some important sense a 'free' agent." (Campbell 1967a, 36) I will outline in more detail later on why I believe this argument prejudges an answer to the problem of free will, which will be referred to as "genuine freedom" from now on, in short: It begs the question. The issue that continuously stirs up the debate is: What actually is the "important sense of 'free' agency?" An answer to this question is neither *ex ante* a compatibilist nor *ex ante* libertarian. Furthermore, the connection between believing in someone's responsibility and punishing them for reasons of reformation and deterrence can indeed diverge. This *normative* problem is, however, not a problem of the theories against which it is directed in the freedom of will debate—namely effect compatibilism and revisionism. This will also be outlined in more detail in this chapter. I mention this version of a "just-desert"-argument in advance to show how closely normative issues of moral responsibility—the expression of praise and blame and the execution of punishment—are connected to the debate about genuine freedom.

Many people suggest that we are ordinarily not concerned about whether we are really free when we attribute responsibility to others and to ourselves (Wolf

1990, 3; Bieri 2013, 431). Only in extraordinary boundary cases, such as when the responsibilities of children, insane people, or drug addicts are questioned, a temporary dissatisfaction with ordinary convictions about moral responsibility arises. But these cases can easily be rectified with a set of criteria that is allegedly unsuspicious and canonical: If someone cannot control one's behavior and act according to reasons (determine ends to be pursued), one is exempt from responsibility and only few adult people do not satisfy these criteria, for example drug addicts and insane people. Apparently, many people live a meaningful life without ever questioning whether their friends and community members are *really* free. Even philosophers have emphasized this point about our ordinary convictions (Bieri 2013, 433; Wolf 2013, 330). Otfried Höffe, for instance, has argued that there is a weaker and a stronger notion of responsibility (Höffe 1993, 22). He believes that only the stronger and more demanding notion of moral responsibility carries the burden of proof for the freedom of will. The weaker version—the freedom to act, which means to volitionally perform actions and being the "doer of one's acts and omissions"—were sufficient in legal contexts and for most of our ordinary activities, including approval or resentment. Furthermore, Höffe argues that radical denegation of responsibility is an irrelevant standpoint. Philosophers, who discuss whether moral responsibility exists at all will do not give a better account of it but perform a futile intellectual exercise without any practical relevance (Höffe 1993, 23). Moral responsibility, according to Höffe, is widely acknowledged. Thus, the debates are instead concentrated on cases of mentally impaired or addicted people, and children, but not responsibilities' general standing or even existence. Visioneers would be clearly subsumed under the umbrella of those people that ordinarily bear responsibility. The baseline of Höffe's reasoning resembles the most common compatibilist convictions, which include pessimism about the possibility of proof of the freedom of will and the belief in the sufficiency of the notion of the freedom to act to distinguish the normal from the "abnormal" agents, which is allegedly all that is required for our *practical concerns*.

For a layperson, this reasoning might be adequate. It is, however, surprising that a philosopher can be satisfied with such approach. Clearly, most philosophical problems do *not* have practical consequences; radical skepticism regarding the external world does not affect our way of doing physics or chemistry (Russell 1999, 111). There are, of course, constant revisions in physical and chemical theories and beliefs. In general, there can be no doubt about the success of these endeavors in uncovering truth. Most of our ordinary intuitions in metaethics and epistemology are modestly realistic; we believe that the sun circles around the earth, and that mass attract each other, and that it is wrong to torture innocent people.[35] Few people ordinarily consider whether the sun *really* circles the earth, whether there *really* is mass attraction, and whether it is *really* wrong to torture innocent people. The pressing questions in these contexts are primarily concerned with the scope of those notions: What are our duties and obligations, how far do they expand and can they affect our integrity, which physical theory should we adopt, and not is there ever a physical theory that represents the world as it is or is there ever moral truth? In short, Höffe's reasoning not only makes a few, but all major philosophical topics futile. For a philosophically minded person this is a result hard to accept (Wolf 2013, 330).

It is too obvious that our ordinary intuitions are incomplete and often incoherent. This is how philosophy usually *starts*: Regarding the freedom of will debate people emphasize, for instance, their belief in the moral responsibility of terrorists. However, when starting to consider that these people were acquainted early with a specific cultural background, specific economic circumstances, and early involvement in extreme religious circles, they start feeling uneasy and realize that everyone is brought up under circumstances beyond one's control. These contingent facts might have a major impact on a person's character and actions. Such philosophical befuddlement and unease soon decrease again, however, the incompleteness and incoherence is already present in such ordinary convictions about responsibility. Shaun Nichols has provided empirical evidence that the folk view on the freedom of will is neither entirely incompatibilist nor entirely compatibilist—but "fractured" and "multifarious." (Nichols 2006, 83; Vargas 2007, 136–138) I will outline in the next chapter that the problem of constitutive moral luck that—amongst other lines of thought— sometimes befuddles ordinary thinking about freedom but does not undermine moral responsibility. The example was

---

35   Both Nagel and Strawson agree on the similarity between radical skepticism regarding epistemology and the freedom of the will problem (Strawson 1985; Nagel 1986). Nagel also expands this skepticism on the problem of the existence of moral truths (ibid., p. 139).

mentioned at this point to support my claim that our ordinary convictions are not as straight and unambiguous as Höffe presumes. On the contrary, such conceptual ambiguities are the sources of many philosophical endeavors, and philosophers readily accept such challenges.

Let me anticipate the following reasoning to support this chapter: The viewpoint developed is largely congruent with the standard libertarian, incompatibilist standpoint. I think it is true that determinism would undermine human freedom. Agency must be conceived as a person's capacity to pursue ends that they determine themselves before acting. People can chose other ends than the one's they pursue and, whatever end they choose, the decision has been up to them. It is unreasonable to conceive agents as mere *hosts* of their motives, reasons, or as the impotent emergent layer of the natural events that occur within and outside of a person's body during their acting. It is also not meaningful to denote the class of events that is from such psychologically determined or event ontological perspective designated as "actions" as actions, because it is not something that has been *done* by a person. Determinism does not permit choosing one's own end and pursuing it. This is because it does not permit anything else from happening other than what actually happens. A simple way to understand this is to imagine an epistemicly perfect being. This being—which is some sort of Laplacian demon—knows all the laws of nature and the initial state of the universe. If determinism were true, such a demon would know what a person in Europe will study in twenty years after graduating from high school, even if that person is not yet born. The demon can universally predict human behavior and, given this perspective, it hardly makes sense to talk about actions in the way we usually do: When we accuse people of not having lived up to moral standards, when we hold them responsible, we account that person for what they did in the light of what else they could have done by, for instance, refraining from what this person did (van Inwagen 1975, 188; Nagel 1986, 121; Wolf 1990, 8). This makes little sense if there was nothing else that they could have done. At every temporal state of the universe, determinism allows for only one pathway into which the future expands. Dismissing the universal predictability of human behavior does not necessarily imply that one also has to dismiss the idea that human behavior is ever predictable. There is a wide range of plausible worldviews that relocate between the extremes of universal determinism and total chaos. Human behavior certainly is largely predictable, but this only proves people's general coherence in thinking and acting. This is a minimal requirement for being considered a person—as a possible candidate of normative evaluation (Williams 1981c). These points should be outlined in more detail now.

The most important pillar of my positive libertarian approach is a commitment to a view according to which the miracle that ought to be explained is determinism and not action. Weaker notions of regularities as, for instance, causal connections without nomological character and nomological regularities of enclosed physical systems are not undermining agency but coexist harmonically with our ability to act voluntarily. The typical question: "How does the freedom of the will fit into a world governed by natural laws?" must be turned upside down to give a proper understanding of agency.[36] Although this viewpoint has often been defended, it is still rather unconventional. Our ordinary language regarding agency is full of causal idioms, which is arguably the source of much confusion (Keil 2000, 1; Verbeek 2011). However, there are three distinct cornerstones that together built a foundation for a plausible theory of free action: First, the notion of the laws of nature (nomological causation), the principle of causation, and universal determinism are notions that need to be carefully distinguished. Only determinism threatens the notion of agency which is incompatibilist stance defended in the following. Second, actions cannot be pressed into causal schemes (as proponents of agent causality and reason causality tried), and third, free agency is properly understood as having the ability to do things instead of doing other things available to the agent. Thus, actions are categories *sui generis* like causes and chance. Actions are *done* for the reasons and when the reasons for which they are done were disclosed to others, the action is justified and made intelligible to fellow people (Wolf 1990, 7). In this manner both reasons, causes, and chance function

---

36   Alfred Ayer writes: "For a man is not thought to be morally responsible for an action that it was not in his power to avoid. But if human behaviour is entirely governed by causal laws, it is not clear how any action that is done could ever have been avoided." (Ayer 2013, 317) The first proposition is perfectly in accordance with my arguments. However, the "if" in the second statement is the big obstacle with which the following chapter is concerned. Ayer poses the question of free will here in its classic form, presuming that the existence of causal laws are the common and undisputable point of departure and agency has to be squeezed into this picture.

as *explanans* for things that go on in the world: Some of these things have an active component (actions), others are merely happening (effects which are caused or events that happened by accident). Through acting, we actualize possibilities that would otherwise "not become true." (Wright 1974, 39) This is mysterious only if we presuppose that the laws of nature must guide anything happening in the world or that agency is intelligible only if we explain it through the scheme of cause and effect. However, causal explanations and explanations through reasons can—with a variety of other ways of making the world intelligible—coexist. In a sense, the notions of the laws of nature and causality are much more mysterious than the notion of agency. This conviction guides my libertarian reasoning in the following sections. I will carry these basic commitments into my discussion of compatibilism, moral luck, and intelligibility. Responsibility (at least in our *judgment* of a person) in any adequate sense presupposes that a person is the kind of agent that will be described and defended.

4.2 Causation and Determinism

The problem of genuine freedom can be parceled into two distinct problems, which are related: The first is determinism, the second is agency. If we reject the theory of determinism, we face the problem of explaining action. What is usually considered the "problem of agency" arises, as I will show, because we try to press agency into a causal corset, and thereby, distinguish artificially the agents of actions from their results. This move raises the problem of agency by underlying an event ontological worldview. Let us start with determinism and distinguish it as clearly as possible from other associated ideas, such as causality and nomological causality. It is remarkable at first that a number of philosophers have questioned whether determinism poses a threat to responsibility before even asking what determinism actually means, and how plausible such theory is (Keil 2007, 15). Some philosophers such as Peter Strawson even claim to belong to the "party of those who do not know what the thesis of determinism is." (Strawson 1962, 1) Therefore, it makes sense to ask: What does the theory of determinism say? Let us first start by exploring some ideas that are related and often conflated with determinism.

First, there is the principle of causation that says everything that happens in the world has a cause (Hartmann 2005, 2). Taking up Kant's terminology we could say that this principle is a perfect example for a synthetic statement a priori: It predicates about *all* events in the world and says that there *exist* events prior to all of these, and the latter causes the former (Keil 2007, 40; Kant 2009, 58). Note that this principle of causation plainly says that there is a cause to each event and not that each cause is a *causally sufficient condition* for these latter events. This no-mological formulation—von Wright calls it the formula of "nomic" causality (Wright 1974, 9)—of the principle of causation is much stronger than the basic formulation of the principle introduced above. The basic formulation is a combination of a universal proposition and an existence proposition and, because of this, the principle of causation is neither verifiable nor falsifiable. Consider that it is plainly impossible to show that *all* events have a cause, but also that it is similarly impossible to prove that there is any event, which does not have one. There can always be causes that remain arcane. However, the principle of causation predicates over events in the world, thus, its status is clearly that of a synthetic statement: It is a synthetic statement a priori. Most philosophers are not willing to accept a principle that is neither verifiable nor falsifiable. Therefore, we should see whether we find a more plausible concept of causation that is less strong in its scope but even stronger in its modal essence to fulfill the nomological condition of causality.

The principle of causation is clearly a proposition about two *events*—a cause and its effect. Take this ordinary example: "P's throwing of the stone broke a window." In this statement, we find two temporally distinct events: "P's throwing of the stone," and "the breaking of the window," mentioned of which the former is in certain contexts adequately put forward as the explanation of the occurrence of the latter. Hence, "P's throwing of the stone" is a cause, and "the breaking of the window" is its effect. If we want to say something general about causation, we need to abstract from this example and understand what makes causation so utterly attractive for the natural sciences. Clearly, this particular causal statement satisfies the "explanatory needs" of the owner of the window, if this person wants to know who of the two boys playing around in her garden broke the window. In science, we are searching for explanations of a more general kind. As Bertrand Russell writes, "[t]he business of science is to find uniformities [...] to which, so far as our experience extends, there are no exceptions." (Russell 1999, 44) Regarding our example, this probably requires finding grounds that supports a statement of the following sort: "Throwing stones at windows makes them break." This is a nomological formulation of the causal principle that asserts the regular conjunction of the two events "throwing a stone" and "breaking a window." If the universal proposition "Throwing stones at windows makes them break" were a true statement, it would be a good basis to predict the events that follow on throwing stones at windows. The particular case "P's throwing of the stone, broke the window" would be logically true, if the universal proposition were also true—since it is an instance of the universal proposition—it can be deduced; whereas concluding to the other direction does not work, the "particular may be true without the general law being true." (Davidson 1980a, 16; Russell 1999, 46)[37] How do we establish the connection between the particular and the universal proposition? In addition, if we can establish a connection, then what is the nature of the connection between these two events? If so, how can it be shown that the events "throwing a stone" and "breaking of a window" are not only regularly but also *necessarily* connected? These are obviously epistemic questions (and not logical ones) (Wright 1974, 50).[38] I have said before that through a great number of our ordinary activities we

---

37   One could distinguish this more clearly by defining causality as nomological causation, while an instance of causation can be a true particular causal statement. I will in the following not employ this distinction.

38   Davidson argues in contrast that "[causal] laws differ from true but nonlawlike generalizations in that their instances confirm them; induction is, therefore, certainly a good way to learn the truth of a law. It does not follow that it is the only way to learn the truth of a law. In any case, in order to know that a singular causal statement is true, it is not necessary to know the truth of a law; it is necessary only to know that some law covering the events at hand exists." (Davidson 1980a, 18) As argued above, I particularly agree with the possibility that we know the truth of singular causal statements without knowing the truth of their universal pendants. However, accidental regularities provide the same grounds for inductive reasoning as nomological regularities. The logical operation of induction obscures the differing natures of those types of regularities because both provide equivalent premises. In

utilize natural courses with which we are familiar: When we plant a tree, we make use of our knowledge of the right time for putting the seed in the ground (spring), how much water to give and how to cut the tree when it starts growing (Janich 1981, 78). We work with and "around" these familiar regularities. Many scientific efforts are directed at uncovering the nature of the regularities that appear to us in our everyday life. Michael Hampe speaks about our pre-scientific experience of regularities using the example of the tides and the changes of season with which we are familiar before we start observing such regularities more systematically (Hampe 2007, 29). Windows break when throwing stones at them, which might in contrast not be one of the regularities with which we are ordinarily (or "naturally") acquainted, but we can start testing whether there is indeed some kind of regularity behind it. We could start throwing stones of difference sizes at windows to see whether they break. We will soon realize that the proposition "throwing stones at windows makes them break" describes a regularity with numerous exceptions. If the stones are thrown with too little force, if they are too small or light, or if the window is too thick, it will not break. Furthermore, if there are birds or other objects that cross the path of the stone, the stone will not even reach the window. In other words: The proposition does not express nomological causality, and if one believes that causality has necessarily a nomological character, then this proposition is not an instance of causality at all.

There are a variety of *ceteris paribus* conditions that need to be met to be able to establish a proposition about the regular character of breaking windows with stones. We could, for instance, repeatedly hit glass surfaces of variable thickness with objects that have variable size and velocity from a variety of angles in a test arrangement. By doing so, we construe an *enclosed system* that differs from the original—the "natural"—situation in significant respects. Thereby, we built an experimental setting that helps shielding the events, which we want to investigate from influences that might affect their conjunction. The situation construed is not

---

each case, we get started on the same qualitative basis of past uniformities. Furthermore, I am afraid that if there is a nomological law that underlies the causal statement "throwing stones at windows makes them break," it will cover a number of completely different, microphysical events. The pressing question will be: Is this law a law about these events, or a completely different law about other microphysical events.

a "natural" situation anymore (in a certain sense, the original situation has not been one either: Windows are no natural kinds). The result of such experiments can be a more refined universal proposition than the one presented above that says, for instance, objects of a certain weight and speed fracture other objects that have a certain density when hitting them from a certain angle. In this case, "to fracture something" is not a very precise scientific concept, however, the baseline of my narrative should be intelligible by now: In solid-state physics and other laboratories sciences, we *construe* experimental situations to gain knowledge about the nomological character of regularities we might have first encountered in nature. Note also that even if physical events are reproducible in enclosed systems, such as technological devices, these systems are vincible and only finitely devoid of external disturbance.

An ordinary television (a Braun tube) reproduces regularities between magnetic fields and cathode rays *only if* it works accurately. When the television breaks the regular conjunction also ends. The view presented here is sometimes called *interventionist* or *experimentalist*, and it is fairly well established in theoretical philosophy now (Wright 1974, 57; Janich 1981, 78; Chalmers 1990, 61). This view regarding nomological causality is also one of the basic commitments of the philosophical school of *methodological culturalism*, which, however, goes in many other aspects beyond the just sketched position (Hartmann and Janich 1996, 54). Regarding the notion of causality, the viewpoint presented here can be roughly summarized as follows: The regularities which are usually called the laws of nature are not to be found "in nature" (Wright 1974, 54). Instead, they are construed in experimental settings. In these enclosed systems, we test the nature and "strength" of the regularities that we have primarily discovered in nature by shielding them from what we considered before as *ceteris paribus* conditions. I think this viewpoint about nomological regularity is fairly plausible. However, I do not have to commit myself entirely to this view. The most important are the following two points emphasized through the previous reasoning. First, we come across regularities, which do *not* have a nomological character: My bus leaves every morning when the bell rings eight times and bakers get up before the sun rises. These are regularities that are *accidently* conjoined and not necessary. They do not denote instances of nomological causation: They do not denote laws of nature. Second, there are types of causes that we ordinarily put forward as explanations of events like the broken window—which are elements of courses that are only *loosely regular*.[39] Straightforwardly we can say, following an established definition, that a regularity is an instance of nomological causality—or in other words a law of nature—if and only if there is an event C (a cause) that is conjoined with an event E (its effect) in such a way that whenever C occurs, E occurs as well. To put this into logical terminology we can say that to be nomologically conjoined the occurrence of C is a sufficient condition for E, while the occurrence of E is a necessary condition for C (Ayer 1952, 55; Keil 2007, 29). Nothing that does not fulfill these conditions deserves the name "law of nature." Laws of nature are expressed in universal propositions such as the one mentioned above: "Throwing stones at windows makes them break." This proposition is formally adequate, but apparently wrong because we can throw stones at windows without breaking them. Therefore, universal propositions of this kind can be falsified by providing a single case in which E did not follow on the occurrence of C. And, as argued above, plausible ways to test the nomological character of regularities, and give evidence that an alleged cause C and its effect E are nomologically and not just accidently conjoined (like the bus start and bell toll), is by attempting to interfere with the course and by reproducing the alleged sufficient condition C and see whether E ever follows.

We are tempted to ask just how many of those invariable laws of nature we actually know. The regularities we instantly recall when thinking of the laws of nature are the seasonal changes and the circadian rhythm. It is immediately obvious, however, that those regularities would require *ceteris paribus* conditions to be fulfilled to meet the conditions for nomological causation: The seasonal changes have uniform character, but we are aware that if the earth were flung out of its current orbit, the seasons will stop occurring and so will the regular rise of the sun in the morning and its setting in the evening—the circadian circle. Both cases obviously preclude interference from humans. Astronomical regularities cannot be pressed into experimental situations. Most (maybe all) astronomical regularities, however, are also not expressed in statements that satisfy the definition

---

39   This statement could be replaced, of course, with a more accurate probabilistic formulation.

of a statement about nomological causality, as introduced above. The statements I have in mind do not describe the causal relation of *events*; they merely describe and relate aspects of planetary movements, such as the masses of these objects, their velocity, rotation, and density in mathematical *formulas*. Mass, forces, density are not events; they are *universals*. Let me discuss this aspect in more detail: The most obvious candidates often brought forward as paradigmatic laws of nature are Newton's laws of motion and the laws of gravitation. Also Bertrand Russell assumes that the most promising candidates for natural laws are those just mentioned and, furthermore, that physicists have been remarkably successful in revealing such laws (Russell 1999, 44). The alleged laws are, for instance, the law of uniform motion: Distance *is* time multiplied with velocity or, formally speaking: $d = v * t$. The "is" in the formula already provides us with a hint of how to interpret the "law of uniform motion" properly. Obviously, this statement does not express a law at all. It expresses a calculable relation of universals, namely velocity, time, and distance for cases of uniform motion. "Distance is time multiplied with velocity," can be interpreted as meaning for example: In case of uniform motion, one can calculate the distance covered by an object by multiplying its velocity and time. Thus, this interpretation suggests that the proposition is rather a working requirement to calculate the distance an object travels in a certain time span.

At this point, one might argue that the "law of uniform motion" might express the relation of universal, but it permits to infer a causal statement about the nomological conjuncture of two *events*. Consider, for instance, the following attempt of such inference: When bringing an object into a uniform motion, which is to be considered an event C, we can observe that the object will be arriving at the end of a path in the calculated time, which can be considered an event E that necessarily follows on C. While the law of uniform motion does not express the nomological character between events, it allows for inference on the nomological relations of events such as exemplarily described in the previous sketch. In the description of this situation, however, we immediately realize that we have already invested a *ceteris paribus* condition namely that someone has to bring the object in question into uniform motion. Uniform motion is hardly, if ever, to be found in nature. Furthermore, if there are stones, or other, bigger objects in the path of the moving object, the event E "arriving at the end of the path" will not take place. The setting must be *isolated* of external disturbances to create a situation in which the loosely regular behavior of objects can be observed. Thus, we are *creating* an experimental situation that is entirely different from any context in which motion can be "naturally" encountered. The "law" of uniform motion is not a law of nature in the above defined sense but rather resembles a mathematical law, a guide to calculate the distance of moving objects with certain uniform velocities under ideal conditions (Hampe 2007, 144). From such laws we cannot infer true universal propositions about the *conjunction of events* (Wright 1974, 60; Keil 2007, 32–33). Knowing the "law" means having a guide at hand that enables us to calculate distances covered by objects in uniform motion under ideal conditions and these conditions can be reproduced and altered in experiments.

As a result to the epistemic challenge of distinguishing regularities that are accidental, or only loosely (with a certain probability) conjoined, and those that are truly nomological, it has been suggested that one must interpret the thesis that the regularities we encounter in nature have sufficient causes, and that they are nomological in nature as a working hypothesis of the natural sciences (Hartmann 2005, 14; Kim 2006, 195; Keil 2007, 41).[40] This understanding is weaker than a realist understanding of nomological regularities is. All philosophical schools dealing with the problem of causality can live with such a formulation: Working hypotheses are there for being tested and refuted at times. Interpreting the thesis "throwing stones at windows makes them break" as a working hypothesis, as the guiding question for scientific investigation is completely legitimate. Understood in this manner, it merely says: "We should test whether throwing stones at windows makes them break," and this *request*—meaningful as it is—is neither true nor false (it might be appropriate or inappropriate, though). Before we turn to the notion of determinism, we should briefly survey two other traditional ways of dealing with the problem of causation and its nomological dimension. First, it has been suggested that nomological causation can be justified by the success of inductive reasoning. Alfred Ayer, for instance, puts forward his own interpretation of Hume's theory of causality and argues as follows:

---

40   Gottfried Seebaß argues that even the heuristic value of such working hypothesis is doubtful because of the recent success of "indeterminate theories" in physics (Seebaß 1993b, 6).

> [...] our view of the nature of causation remains substantially the same as [Hume's]. And we agree with him that there can be no other justification for inductive reasoning than its success in practice, while insisting more strongly than he did that no better justification is required. For it is his failure to make this second point clear that has given his views the air of paradox, which has caused them to be so much undervalued and misunderstood. (Ayer 1952, 55)

Since we are not primarily interested in historical analysis here, we can skip the question whether Hume really promoted such position. We take Ayer's view that our success in inductive reasoning is evidence enough to provide a solid ground for causation. Clearly, in many practical contexts, we make use of arguments containing premises about previously observed regularities, and we generalize those regularities and utilize such knowledge. We could not make use of natural courses in the way described above had we not at least superficial insight in their regular working mechanisms (see also section 3.5). To successfully plant a tree, we need an understanding of the natural course that is that (most) trees do not grow without water and sun. It is, however, obvious that watering a tree and planting it in a sunny spot is not sufficient to make it grow.

Hence, the regularities—even if they are understood as instances of causation—do not have to have nomological character. In addition, my predictive success when forecasting the start of my bus in the morning will remain astonishing and accurate as long as the bus company does not change the schedule. Accidental regularities provide the same basis for inductive reasoning as real nomological regularities. Prima facie, it does not seem that predictive success gives us the definite clue for distinguishing nomological from accidental regularities. Again, we are at a loss regarding this issue if we do not begin shielding the regularities we want to investigate and thereby "denaturalize" them. The second approach to be briefly discussed is also associated with Hume (this time possibly with more evidence from his original writings).[41] It is the thesis that we acquire a notion of causality through habituation. Bertrand Russell writes in this manner: "Experience

---

41 The following passage from Hume's Treatise might substantiate this interpretation: "'A cause is an object prec'dent and contiguous to another, and so united with it,

has shown us that, hitherto, the frequent repetition of some uniform succession or coexistence has been a *cause* of our expecting the same succession or coexistence on the next occasion." (Russell 1999, 43) When becoming familiar with typical regularities encountered in our everyday life as, for instance, the uniform sensation when drinking peppermint tea, we start associating certain events (the alleged causes) with other events (their alleged effects). In other words, we develop an expectation regarding these regularities. We start *expecting* that drinking peppermint tea will produce the same taste sensations as ever before. We can call this theory "the habit of induction"-theory of causation. In this case, it is one thing to know why we *believe* in the continuation of a natural course (because we have observed a regular conjunction before), another is to provide evidence for believing that the "nature" of this very conjunction is necessary, and that it is essentially different from accidental regularities (Russell 1999, 43). Since there is a difference between them, the reliance on "the habit of induction"-theory is futile for demarcation: Both types of regularities will produce an expectation in the human mind and make us anticipate certain effects on given events (Chalmers 2007, 172). The "habit of induction"-theory is a psychological theory about *expectations' development*. It is not a theory about the nature of causality. Moreover, it actually presumes a (vague) idea of causation: Whether the "habit of induction"-theory is a true theory about the regular conjunction of experience and expectation is itself an epistemic question. So far, we have outlined the difficulties of causation and its nomological variant in great detail. We have to ask at this point how the notion of determinism is related to this and what it actually means.

Determinism must be interpreted as a thesis about a particular type of causation; more specifically, about nomological causation. Moreover, in contrast to particular nomological regularities, universal determinism is a thesis about the whole

---

that the idea of the one determines the mind to form the idea of the other, and the impression of the one to form a more lively idea of the other.' [...] Again, when I consider the influence of this constant conjunction, I perceive, that such relation can never be an object of reasoning, and can never operate upon the mind, but by means of custom, which determines the imagination to make a transition from the idea of one object to that of its usual attendant [own emphasis] [...]." (Hume 2003, 122) Because of this and similar remarks Keil and Hampe consider Hume as a "nominalist" regarding causality (Keil 2007, 157–160; Hampe 2007, 75).

world and *all* the events it entails. These events are conceived as being necessarily conjoined as causes and effects. Thus, according to a traditional view, determinism says that "[the] world is governed by (or is under the sway of) determinism if and only if, given a specified way things are at a time t, the way things go thereafter is fixed as a **matter of natural law** [own emphasis]." (Hoefer 2016) This seems to be a proper definition of determinism with which a fairly large number of philosophers start their discussions (Keil 2007, 35). Central in this definition is the concept "natural laws," which has previously been defined as denoting a nomological connection between certain causes and effects. In pre-modern times the modal source of the necessary regularities was conceived as an omniscient God (Hampe 2007, 72). To be more precise: In pre-modern times (at least) two different images prevail: First, the antique idea of the orderliness of the cosmos and, second, the Christian idea of eschatology. Both ideas are in important respects different to the above outlined conception of nomological causation. The first idea of cosmic orderliness does not imply the necessary conjunction of events. The second—the eschatological view—does not even utilize the notion of regularities. It is a view that expresses that the course of the world eventually leads to a final state: the end of time after the apocalypse. This view is *teleological* and not causal. By distinguishing the idea of the cosmic order and eschatology from determinism, we begin to encounter a bandwidth of possible (and coherent) alternative worldviews. Clearly, this does not prove determinism to be false, but it extends our perspective on the world. By contrasting these positions, we also realize that in opposit to the teleological view, which regards God as the modal source, the power that pushes its inescapable moving towards the end of days, the burden of providing the modal source of determinism rests entirely on the notion of the laws of nature.

Before we critically assess the idea of determinism, let us briefly recall again in our own words what it says. We can say that a state of the world at a given time is a class consisting of all events in the universe occurring at that time. Given the nomological character of determinism, we can say that this class containing all events at a given time is the sufficient cause for the successive world state that constitutes a class containing all its effects. Assuming the previously outlined definition of nomological regularity, the world state C must be a sufficient condition for the occurrence of the world state E. How can such view be defended? One way to defend the view would be to assert that we have already detected a number of nomological regularities in the natural sciences and extrapolate from the existence of these regularities to all the other events in the world. Even if it were true that we already had discovered a number of nomological regularities—an assumption that has been critically assessed before with sobering results—this type of inductive reasoning is as problematic as any other instance of inductive reasoning (Sand and Jongsma 2016b). It might be plausible, but clearly does not follow that *all* events are the results of the laws of nature from the propositions that *some* events are the results of the laws of nature. Another positive way would be to verify that all events in the world have sufficient causes. This is obviously practically impossible and given the difficulties of distinguishing accidental and nomological regularities outlined above, verificationism seems to be in general in a defensive position; Verifying instances of cause and effect occur regularly in both accidental and nomological instances of regularity. Let us, therefore, ask it the other way around: What would it take to refute or to falsify the determinist worldview? Consider what Mackie writes about this issue:

> Then the determinist thesis is that for **every event there is an antecedent sufficient cause**, that is, a temporally prior set of occurrences and conditions which is sufficient, in accordance with some regularity, for just such an event and which leads to it by a qualitatively continuous process. This thesis **would be falsified if there were two antecedent situations which were alike in all relevant respects, but had different outcomes.** We make progress towards confirming it in so far as we find what appear to be satisfactory causal explanations of more and more kinds of occurrences. It is an empirical thesis, which only the progress of science will either gradually confirm or more dramatically disconfirm [own emphasis]. (Mackie 1990, 216)

The quote shows that Mackie agrees with the definition of nomological causality which regards causes as antecedent sufficient conditions as outlined above. He believes, however, that determinism would be falsified if there were two antecedent situations, which were alike in all relevant respects but had different outcomes and, related to this, that determinism is an *empirical thesis*, which can only be disconfirmed by science. Given the foregoing reasoning, we can agree that particular candidates for nomological causality can be falsified by showing that an event is not a sufficient cause of an effect; thus, the effect does not necessarily succeed on the event. This, however, only falsifies the *particular* alleged causal relation; it does not falsify the *universal* thesis of determinism. As von Wright writes: "The problematic proposition [of determinism] is that *all* changes are of this [nomological] nature." (Wright 1974, 106) Given the above outlined description of determinism, it remains, for instance, possible that there are plenty of other candidates in the class of events that could be sufficient causes of the events in question. Note that in the definition given above, it was not said that it is *ex ante* clear which events are the sufficient causes of the events in question. If we falsify *one* of the candidates for nomological regularities, we do not falsify the thesis of determinism *as such*. The class of events that precede the class of alleged causes might contain other events, which might have been the sufficient causal conditions. A determinist could argue that we have just investigated the wrong connection—we sleuthed the wrong clue.[42] From this perspective, the thesis of determinism is not falsifiable, just like the principle of causation mentioned before. There are always events not yet considered that might be the sufficient causes sought after. Another understanding of determinism differing from the one presented above conceives the states of the world at a point in time rather than as classes of events, but as events in themselves.[43] These events are extremely rich and maybe containing billions of physical and microphysical sub-events. Clearly, such ontology of supra-events is dubious. It establishes an event that has never played a role in any kind of explanation of natural phenomena that attract our scientific curiosity. Furthermore, even if we accepted such supra-events in our ontology, we were in no better position in defending determinism. Total states of the world cannot be observed, let alone being tested for having nomological conjectures with their successors and this is a

---

42   This could be even more complicated, if we accept that causes can operate "at a distance" as von Wright writes, meaning that the events in the successive class might be caused by events that do not even occur at the same point in time as the class containing the alleged causes but (long) before: "If the answer to the question [of distant causation] is affirmative, it is not enough to know just one pair of successive states (and the laws) in order to be able to predict all future changes. For some of the changes which take place in the future may, in fact, be the effect of changes which were anterior to the given pair of successive states." (Wright 1974, 119)

43   In the Laplacian version, it is not clear whether he prefers this or the other reading. Laplace speaks of world states, but not of their composure (Laplace 1932, 1).

necessary condition in a determinist theory. It should become clear by now: The thesis of determinism is a metaphysical thesis *par excellence* which can neither be supported nor refuted by the natural sciences in the manner Mackie suggested (Keil 2007, 38; Sand and Jongsma 2016b). Thus, Patrick Suppes summarizes rightly:

> The metaphysics of either determinism or indeterminism is transcendental, in the sense that any general thesis about the nature of the universe must transcend available scientific facts and theories by a wide mark. (Suppes 1993, 254, 1994, 454)

Moritz Schlick and many other authors support this viewpoint. Schlick claims that no one knows whether determinism is true or not: "Whether, indeed, the principle of causality holds universally, whether, that is, determinism is true, we do not know; no one knows." (Schlick 1962c, 144; Wright 1974, 136; Hartmann 1996, 74) As a last resort, determinists might find shelter in a formulation that is weaker than a verifiable or falsifiable principle. They could propose to settle by conceiving determinism as a *plausible* scientific worldview. Most of us, for instance, can neither falsify nor verify that they have eaten something on Christmas Eve some ten years ago. Still it is plausible to assume that we all ate something on that evening. If we had an anchor that is a true belief, which we can then sufficiently relate to the idea of determinism in the way just described, we could at least establish determinism as being a plausible worldview. Such anchor could have been the laws of nature. If, as Russell asserted, the sciences had been tremendously successful in uncovering the nomological connections in the regularities we encounter in nature, we had found our anchor. We could then suggest that it is likely that science will continue making progress in uncovering those laws. This success would be a ground to believe in determinism's plausibility. But how successful has science really been in this respect? Our search for a paradigmatic nomological regularity has been sobering to say the least. The most promising candidates, such as the law of uniform motion and the law of gravitation are no laws at all and other regularities satisfy the nomological condition only if shielded from external influences.

Let me summarize briefly: We distinguished three notions of causality. The principle of causality that says everything happening in the world has a cause. This formulation does not have nomological implications and is neither falsifiable nor verifiable. We distinguished nomological and non-nomological causality and defined nomological causality as the necessary connection between events so that the occurrence of antecedent causes are sufficient conditions for the occurrence of their effects. When speaking about causation we are often not aware of such nomological connections and still put causal statements forward as explanations ("P's throwing of the stone broke the window"). In order to get certainty about the nomological character of regularities, we cannot do so without interfering into "natural" courses. Determinism properly understood relies on the laws of nature—another form of nomological causation—and extrapolates their scope over all events in the universe (Seebaß 1993b, 3). This is also neither falsifiable nor verifiable and given the lack of success in finding any laws of nature, hardly a plausible theory. Let us now turn to agency, which seems to entail problems on its own.

4.3   Agency and Responsibility

Thus far, we have explored what determinism means and how it can be distinguished from other (weaker) notions of causality. While many philosophers believe that determinism can neither be proven right nor wrong, the reasoning provided before strongly suggests, in contrast, its implausibility. That does not mean, however, that the question whether it would threaten the understanding of ourselves as responsible agents is meaningless. On the contrary, since there is a large number of challenging compatibilist theories based on the idea that we can neither make sense of agency in a determined nor in an undetermined world, and since those alternatives are suggested to be exhausting, we should settle with a concept of agency that is compatible with determinism—as the familiar argument goes. Such reasoning has found enough adherents to be discussed in its own right. After reviewing the possibility of determinism, Mackie writes that "[the] arguments for and against determinism about human actions, then, are inconclusive, so we must turn to the hypothetical question whether, if such determinism holds, it significantly undermines our moral ideas, so that if we accepted determinism we should, for consistency, have to make radical changes in our notions of choice, responsibility, credit and blame, resentment and gratitude, and perhaps in even more central moral notions like those of goodness, justice, and obligation." (Mackie 1990, 220) Mackie gives a typically compatibilist answer to this hypothetical question.

He argues that acting intentionally is sufficient to be held responsible. In contrast to Mackie, I will argue that determinism undermines responsibility. Responsibility is attributed to persons for their *choosing* of certain ends *in the light of other ends they could have chosen* (Wolf 1990, 8; Bieri 2013, 45). Determinism is a theory that describes the world as a succession of world states nomologically conjoined— it describes what is *happening*. Regarding such worldview *active* and *open-ended* deliberation of ends, the formation of intentions, active deciding and doing has no place. In contrast to Mackie's suggestion, plainly the bearing of intentions cannot be a sufficient condition for being responsible. These intentions must have been freely chosen and this choosing freely is also a necessary condition. Before we come back to compatibilism and Mackie's reasoning later on in this chapter, let us first explore the hypothetical question whether determinism—implausible as it is—would threaten responsibility. This question is a proper vantage point for finding an acceptable definition of agency. Free agency does not automatically become intelligible, if determinism is shown to be implausible and our ordinary notion of agency shown to presume the falsehood of determinism. It could still be possible that agency is an unintelligible idea. As mentioned before, I believe this appears to be the case only when we try to press it into a causal scheme.

When we ascribe responsibility, we ascribe it for things that have been done by persons. Things people do can be demarcated from mere happenings or behavior so to speak, by virtue of having been intended based on foregoing deliberation. Thus, a person who intends to murder the King of Denmark can fail in pursuing this end. The realization of her intention can fail when the pursued end or the intended goal does not occur—when the King of Denmark continues living. On the contrary, people who stumble down the stairs and break a light switch, thereby producing an electrical fire, cannot be said to have been successful or unsuccessful in what they did, because they did not do anything intentionally. What has happened was not directed at an end and not actively pursued. Let us consider a typical example of deliberation in order to understand how decisions are made and actions performed. Take Carla, a high school graduate: She has just finished her studies and is transitioning into a college degree program (Kane 2007, 2; Sand and Jongsma 2016b). She considers law and psychology as possible majors and collects information about a number of universities in the United States. She starts thinking about the subjects she liked most in high school, remembering for which subjects she received the best grades, how far she would be away from her parents and friends, how many fees she would have to pay and how well the job opportunities are for each of these fields of study. She considers all this and decides on law at the University of Michigan: Michigan has a solid reputation, is not too far from her home in Illinois, and affordable. She applies and gets a placement. During this process new reasons towards her decision might emerge due to shifts in her social environment (she gets to know a new partner), and meanwhile, she realizes that other reasons became obsolete (the friends to whom she wanted to stay close have decided to move too). She can approach this process systematically, for instance, by drawing a table with columns for the different majors and universities and subsume in each column the pros and cons. In any case, even if it is not approached in such a highly systematic way, the inscription at a university is a conscious process. Inscribing is a conscious activity consisting of a number of steps, presuming the formation of an intention. One of the crucial aspects of this narrative is that she had a number of different alternatives. There are numerous ways in which she can approach (and structure) her decision-making process, and eventually she can chose where and what to study. We can call this, following the traditional literature on the free will problem, the condition of *alternative possibilities* (AP) (Kane 2007, 14). Carla, so it seems, has a number of alternative possibilities and her struggle to decide concerns finding the one that is most suitable to her. Given that the previously discussed notion of determinism (both in the version of world states as classes of events and world states as events) says that everything that happens, happens because of its past and the nomological conjecture between this past and its successive states, it is impossible that anything else could have happened. Past events are always sufficient causes for future events, and there is only one possible course, one possible state of the world succeeding from another. The succession of events in the universe (or the states of the universe) is fixated from its very beginning by virtue of the modal character of the natural laws, therefore allowing only *one possible future at every point of time* (Seebaß 1993b, 4; Kane 2007, 2; Keil 2009, 35). Since this is the case, it would hardly make sense to say that Carla could have chosen to inscribe in Yale instead of inscribing in Michigan. A future in which she inscribes for Yale is an impossible course in a deterministic world: There is always only one future, and the agent cannot make a change to bring it about. Yale was her second preferred choice, but she considered it as being too far from home. There are no external constraints regarding this decision. She had the required grades for each university, their reputation is (we suppose) fairly the same, and the fees are equally high. Her parents support any decision she makes. If she inscribes for Michigan, it might appear to her that it is the result of her willed action: She will be happy that she finally decided and if her studies turn out extremely well, she might one day proudly refer to this as the best decision of her life—which is for most of us a common theme regarding such impactful choices. Regarding its implications for responsibility, such theme is much more suspicious than we ordinarily think when we encounter it. But does that—the pride, the happiness for having become clear about one wish—make sense, if nothing else could have ever happened? It seems that the whole point of deliberating rests on the openness of the future (Bieri 2013, 45). The meaning of the act of deliberation rests on the fact that it is not clear beforehand which major she inscribes for: That is the reason why she first has to find out what she wants.

There is a classic counterargument to this reasoning, which should be briefly discussed. This argument concerns the meaningfulness in the act of deliberation. Clearly, there is a difference between the truth of determinism as the necessary succession of states of the world, and knowledge about this truth and about which state of the world will factually occur. The first (truth of determinism) can be called the "ontic" thesis of determinism; the second could be called the "epistemic" thesis of determinism (knowledge of the laws of nature and the predetermined future) (Wright 1974, 120). Given that Carla does not have insight into the laws of nature that built the necessary connections between the past and the future, she does not know what she will inscribe for—she does not have the complete epistemic picture of her determinedness although she knows a few regular courses of which she makes use of within her action (the duration of the mail delivery service, for instance). Given her "epistemic incompleteness," deliberating makes sense. She does not know at which University she will inscribe, and since this is the case, she could meaningfully continue deliberating, while at the same believing that it is predetermined what she chooses. In contrast, the basic fatalist mistakes rests on the conflation of those dimensions—the ontic and the epistemic dimension of determinism. When the fatalist has an illness, they will, for instance, believe that they will die no matter what they do and, therefore, decide to stay home and face death rather than go to the doctor (Seebaß 1993b, 13; Meyer 1999, 253). In fact, it *might* be true that they will die one day or even now if ontic determinism is true (she most definitely will one day, even if determinism is not true). But where does the certainty come from that the doctor cannot help her now—that they have to die *this time*? Such knowledge is not implied in the ontic thesis of determinism. It is one thing to know that one day you have to die; another thing is to know when it will happen (see also my thoughts on phlegmatism in chapter two, and (Seebaß 1993b, 4; Pauen 2004, 33; Keil 2007, 20–21)). Our epistemic imperfection builds the ground for believing that deliberating about what to do in a given situation makes *sense* even if it is predetermined—because we do not know just *how* it is predetermined. Through this distinction, the determinist can uphold the belief in the meaningfulness of our deliberations. In the Laplacian version of determinism, both the epistemic and the ontic dimension are intertwined. In the Laplacian world there is only one possible world state succeeding the others and the Laplacian demon (an omniscient intelligence as he calls it) knows which one is next (Laplace 1932, 1). We are obviously not compound like a Laplacian demon; our epistemic resources are strictly limited. However, imagining such a being helps to understand the question that concerns human freedom. We have asked before and can do so now under the impression of the idea of the Laplacian demon. Is it meaningful to say that Carla *chose* to inscribe if only one thing could have happened namely that she inscribes for Michigan? If there were only one possible future, why would we say she has *chosen* it? Would it then not be more appropriate to say that the inscription in Michigan merely happened like anything else in the world? I think it would then be appropriate to say that her inscription happened just like the eruption of a volcano or other natural events that can be causally explained. This has to do with the kind of theory determinism is—a causal theory. A causal theory explains things that happen, but inscribing at the University is something that is done. Let me outline this crucial opposition in more detail.

It is not meaningful to say that Carla *did* anything at all if determinism were true. The reason is not primarily a conflict between determinism and AP, although determinism makes other pathways to the future than those that really occur factually impossible. The primary reason is something most adequately denoted as the *authorship condition* (AC) of agency.[44] AC seems to imply AP. AC says that people do things for certain reasons. Doing things for reasons always implies that one does not omit doing these things, and this is always a possible alternative that the doer has, which is AP. AC conflicts with determinism and other causal theories of agency for one simple reason: Causal theories are *kinds of theories* that explain the occurrence of events by naming their antecedent states and relating them either nomologically or in other (weaker) ways to their effects. These kinds of theories provide us with explanations of why things *happen*. Such theories, however, are entirely inadequate to explain or make intelligible things that are *done* (Keil 2000). The condition of authorship says that agents do things for specific reasons, which implies—doing things even means—the author is *active*, and she actualizes possibilities that would otherwise not occur. The problem with determinism is, thus, not so much that it would undermine freedom; the problem is that it would explain freedom in the sense of agency away. The same applies to causal theories of agency. Agency is often conceived as implying a kind of control. Speaking of agents as having a "special" power to do what they want, if not coerced, or controlling their bodily movements is misleading because it already evokes causal associations. Confusingly, we ordinarily apply the language game of "control" and "power" to, for instance, traffic lights that "control" traffic jams, to electrical fuses that "control" the energy regulation in a power supply system and this might have stimulated attempts to conceive actions (or their reasons) as causes. But only

---

44   Kane calls this condition the "ultimate responsibility" condition (UR), while others speak of "sourcehood" (Urheberschaft) (Nagel 1991b; Wolf 1990, 41). Kane writes: "Free will also seems to require that sources or origins of our actions lie 'in us' rather than in something else (such as the decrees of fate, the foreordaining acts of God, or antecedent causes and laws of nature) outside us and beyond our control. I call this second requirement for free will the condition of Ultimate Responsibility (or UR, for short); and I think it is even more important to free will debates than AP, or alternative possibilities. The basic idea of UR is this: To be ultimately responsible for an action, an agent must be responsible for anything that is a sufficient cause or motive for the action's occurring." (Kane 2007, 14) I think that Kane does not do the libertarian theory a great favor when calling the occurrence of an action as originating from sufficient causes. In addition, the condition of UR raises associations with conceptions of agents that are able to make themselves from scratch.

agents do things, and doing things has an active component. This active component is the key to understanding agency and authorship (Bieri 2013, 31). In short: Things that are done (actions) and things that happen (events) constitute entirely different *kinds*. It is, therefore, also a categorical mistake to believe that one could explain actions with causal theories. Regarding this reasoning, we are finally at the heart of the question of determinism. We can now ask in the opposite direction than traditionally done: How can a worldview such as determinism be true that suggests that there is a comprehensive description of the world in terms of events and their causes while there are such things as actions? If determinism were true, not only were there no alternative possibilities, there were neither agents nor actions. But since there are agents and actions, determinism cannot be a true theory that gives a complete description of what is going on in the world.

The problem of action is not solved yet. If we explain the fall of Pompeii by saying that the Vesuvius erupted and destroyed it, we denominate the cause of this catastrophe and by doing so we also provide an explanation of the event. The eruption of the Vesuvius *explains* the fall of Pompeii. What is the equivalent to this kind of (non-nomological) causal explanation in the realm of actions? How do we make actions intelligible, and what are they if not processes? As mentioned before, our ordinary language is unfortunately extremely imprecise in this respect. On the one hand we address events in which neither actions nor agents have been involved with terms from the normative "language game" (Keil 2000, 1; Nida-Rümelin 2011, 19). One can, for instance, say that the broken V-belt has *been responsible* and that the car broke down or that the bad weather *is responsible* for the delay of the train. We also use, as mentioned before, verbs to describe the processes of inanimate things that definitely cannot act: We say that my computer *wants* to annoy me today, or that the dessert *threatens* the life of its inhabitants. On the other hand, we happen to address agents or their actions as causes—the present research has so far not been free of such verbal "confusion." We say, for instance, that a person induced a danger, caused an accident, produced a conflict or their action brought about a dangerous situation. Here, our ordinary language is not as straight as we would prefer. There is a terminological overlap of causal terms applied to agents and their activities and intentional language applied to purely causal, inanimate processes (Verbeek 2011). We do not have to clear our ordinary language of those confusions—they are in the source of a number of thoughtful and evocative metaphors in literature—but we should get clear what we mean when we speak about actions and when we speak about (non-intentional, passive) processes. It is likely that this is the origin of many philosophers' motivation to impose causal language on actions. The two most sophisticated versions of such causal theories are probably the event causal theories of Donald Davidson and the agent causal type of Roderick Chisholm and Richard Taylor (Davidson 1980a; Chisholm 2007; Taylor 2013). Let me just sketch their basic commitments without getting into too many details. For this purpose, we are merely interested in what commonly distinguishes these approaches from my own. In *Actions, Reasons and Causes* Davidson first established a theory that conceives reasons as causes (Davidson 1980b, 12). In his article *Agency* he writes that "an important way of justifying an attributing of agency is by showing that some event was caused by something the agent **did** [own emphasis]." (Davidson 1980b, 48) Here it seems that the *action* (and not the *reason*, as suggested in *Actions, Reasons and Causes*) is conceived as a cause. Before this passage, he argues closer to the original theory in *Actions, Reasons and Causes*: "With respect to causation, there is a rough symmetry between intention and agency. If I say that Smith set the house on fire in order to collect the insurance, I explain his action, in part, by giving one of its causes, namely his desire to collect the insurance." (Davidson 1980b, 47) Here, the view that sees *reasons* as the causes of actions becomes more obvious. Smith's intention to collect the insurance is the cause for burning down the house. In contrast to such event causal theory, Chisholm and Taylor conceive the *agents* as causes.[45] Chisholm writes, "[...] if a man is responsible for a particular deed, then [...] there is some event, or set of events, that is caused, not by other events or states of affairs, but

---

45   In his Grounworks on the Metaphysics of Morals Kant conceives the will (and rationality) as a type of causality (Kant 1996, 81). He understands rational beings as partaking both in the world of appearances ("Sinnenwelt") and in the world of understanding ("Verstandeswelt") (Kant 1996, 88–91). Kant suggests an ontological distinction between those two worlds (Schönecker and Wood 2002, 206). Kant's conception provokes the typical inter-actionist-explanatory issues, which will be discussed in the next section. How does rationality cause events in the sensual world that are completely governed by natural laws? In contrast, the distinction between actions and causes entertained in the present argumentation is derived from a prerogative of the methodological before the ontological.

by the agent, whatever he may be." (Chisholm 2007, 51–52). In a similar fashion, Richard Taylor writes, "it [Taylor's theory of agency] involves an extraordinary conception of causation, according to which something that is not an event [the agent] can nevertheless bring about an event [...]." (Taylor 2013, 310) I also do not think that the conception of agency is so extraordinary, nor do I believe that we have to or even should conceptualize it as form of causation. A number of highly technical critiques against both of these approaches have been brought forward, which I do not intend to discuss here (Alvarez and Hyman 1998). I think that each of the just mentioned approaches fail for the same reason that determinism would fail in providing a comprehensive worldview. They explain what is happening in terms of events and their causal connections, thus leaving out the crucial aspects of agency: The doing of things (Keil 2000, 476). Thomas Nagel puts this failure concisely in the following passage from his *Subjective and Objective*:

> The recent attempts to analyze action in terms of agent causation [Chisholm] rather than event causation [Davidson] is instructive because it reveals the true source of discomfort with determinism. The problem is that when one views an action as an event causally connected with other events, there is no room in the picture for someone's *doing* it. But it turns out that there is no room for someone's *doing* it if it is an event causally *un*connected with other events, either [which is the problem of chance]. [...] It is a doomed attempt to capture the *doing* of the action in a new kind of causation. (Nagel 1991c, 198)

Here, Nagel establishes the two extremes of the subjective (internal point of view) and the objective (event ontological point of view) through which his arguments concerning free will and responsibility meander in his articles and books. In both notions agency does not seem to have a proper place: On the first view because the authorship aspect of agency is eliminated, on the second because the actions look like the results of chance. Those viewpoints are presented as being exhaustive, however, they are not; besides events, the world also contains agents and their actions. Nagel establishes event ontology as a comprehensive objective worldview, but does not provide us with an argument as to why this view should be compelling. As argued before, if the event ontological view is conceived as a propositional statement about the nomological succession of world states it is a priori false because of the existence of actions. If it is merely conceptualized as a perspective (as Nagel does it in other writings) it might *seem* incompatible with other worldviews, but then—as a perspective—it is also not truth-functional.[46] However, Nagel correctly expresses here that this is a *categorical mistake* to conceive the agent, his reasons or his actions as causes. At the end of *Actions, Reasons and Causes* Davidson asks astonished—reacting to a remark of Abraham Melden—"[why] on earth should a cause turn an action into a mere happening and a person into a helpless victim?" (Davidson 1980a, 19) This is a typical compatibilist response because determinism would, if it were true, not be coercive in the sense in which social laws are coercive. In fact, the causal description and the determinist theory do not undermine agency—as Davidson rightly argues here—these descriptions just do not capture the actual phenomenon, the authorship aspect of agency. The causal description does not turn the action into a happening; it tries fallaciously to describe it as one.

4.4   Explaining Action

What are actions and how can they be explained? Actions are a category *sui generis* like causation and dispositions—they coexist with those, and we have our own way of making their working mechanism intelligible (Alvarez and Hyman 1998, 233). Actions are what people do for reasons, as described with the example of Carla (Sand and Jongsma 2016b). Agents become clear about what they want through deliberation, and from an external point of view we can *rationalize* the decision making by asking for the *reasons* the agent had. By disclosing her reasons to others, Carla provides *a sort of explanation* that forms the pendant to the causal explanations mentioned above. When we speak to her and she says that Yale has just been too far away from home (given that Yale has been analogous in all other relevant respects), this datum makes her decision and action intelligible to us. In cases of moral concern, the reason can excuse her or make her decision look even worse. Not all reasons are (morally) good reasons (Keil 2007, 142). Causal theorists will be tempted to ask at his point: Is there not also an event (or even more

---

46   This will be outlined in more detail in chapter five.

than just one event) happening when she inscribes to the University of Michigan? And cannot this event be physically described because it has physical components: It leaves a visible trace, for instance, in the form of her signature in the yearbook of the University, in the form of the printed documents of her certificates in the Universities' mailbox, and finally her (physical) body appearing in one of the University's hallway at the beginning of the semester. How is *she* related to these events, if not as a cause? How do we explain these *events*? My response is twofold. First the actions entail events *conceptually.* We do not need to divide the action into distinct entities (the agent, the action and the event) and then ask for their (causal) relation to make this mechanism intelligible: Saying that "P murdered Q" *implies* Q's death. When P says that he murdered Q because he wanted to become the new King, this is the explanation for the occurrence of Q's death. Moreover, we can also employ causal language to those events that happen simultaneously to the action. By doing so, it is crucial to keep in mind the vast deficits of our current causal knowledge mentioned above. And, if assuming for the sake of argument that we had a complete causal knowledge of the events underlying an action, it is crucial to remain aware that we still would not provide a description of the action, but rather of the events happening simultaneously. Let me explain both of these points in more detail by contrasting my reasoning to von Wright's analysis of agency. There are many similarities between our approaches but also some differences. Those differences will help me to put my theory in a nutshell. Von Wright writes:

> Causal relations exist between natural events, not between agents and events. When by doing p we bring about q, it is the happening of p which causes q to come. And p has this effect quite independently of whether it happens as a result of action or not. The causal relation is between p and q. The relation between the agent and the cause is different. The agent is not "cause of the cause," but the cause p is the result of the agent's action. The effect q is a consequence of the action. The relation between the result and the action is intrinsic. The result must be there, if we are to say correctly that the action has been performed. The existence of specific causal relations, and the operation of causal factors, is thus independent of agency and of the interference of agents with nature. One could express this by saying that causation is ontologically independent of agency. (Wright 1974, 49)

Von Wright expresses here precisely my worries with the causal theory. Causality holds between events and not between agents and events. However, then he writes confusingly that "doing p," or the "happening of p" causes q. In this description lies the crucial difficulty of his analysis, as I believe. It would be more precise to say that there is a description of the action as the causal source of a certain event, but that this description is not the description of the action. The action is adequately described as the doing of a person for reasons. "P murdered Q, because he wanted to become the King of Denmark" implies (if is a true statement) that Q is dead, which is an event. We can say that Q's death is the result of the action, and the consequence of the result is that Denmark has to search for a new King. There can be a causal description between the result of the action and its consequences, which is comprehensive because there are no further actions involved. We can also give a causal description of the sub-events happening simultaneously while P murders Q. We can, for instance, describe that the stab wound in Q's belly has caused loss of blood which was eventually fatal. This is a (non-nomological) causal description of the event "knife penetrates body" and the occurrence of the event "loss of blood." These events are causally related.

But this description does not describe the action; it describes the relation of two events that happen simultaneously. The only proper description of the action is: "P murdered Q" and the explanation is that he wanted to become the new King of Denmark. In cases of normative concern (which a murder most definitely arouses), the explanation also functions as justification. Consider another example: Cooking means boiling water, putting food in it and waiting until it is cooked through. It is utterly misleading to ask: How does "cooking" cause the water to boil and the food to become edible? By asking such a question, the causal theorist imposes an artificial distinction into the action; a distinction between the actions and the events they entail. The source of confusion seems to be an "eventification" of the concept "cooking." Cooking *means* boiling water, adding foods to it and waiting until it is boiled thoroughly. "Cooking" is not a separate event that causes other events. We can describe the chemical mechanisms that happen during the heating process in the food and maybe some of them are causal, but then we are not describing the cooking but those chemical mechanisms. In the same manner: acting *means* doing things for reasons and doings things *means* that there are certain events happening (at least when the doing is not interfered by anything else, and what has been done is the *trying* of doing something which is also a doing). Human freedom consists in the ability to do things and this is mysterious only if one tries to press it into a description that conceives the agent or her actions as events and search for an explanation that goes beyond the reasons an agent had for performing it. In this manner the ability to suspend an initial preference—which has been mentioned before as a condition for freedom—is just *another* action which agents are capable of performing (Keil 2007, 133; Locke 2008, 142). We can continue deliberating instead of performing an action of whose goodness we are not entirely convinced, just as we can murder people, cook pasta, and throw balls at windows. Moreover, we can meaningfully demand from people to do such things because (most of the time) they *can* do them—although that is not in itself a good reason to demand them. When we do those things intentionally, we are also responsible for them because we could have set ourselves up for other paths and performed other actions. This is the meaning of agency and if this meaning is properly understood, there is no further problem of freedom. As mentioned before, the problem of freedom is often misunderstood beforehand as the mystery of causal interference. But agency is neither causal, nor mysterious.

The view presented here can be broadly considered as a non-naturalist theory of agency. It has been challenged by a number of naturalistic inclined philosophers. Before I finish this section, I will briefly discuss a recent and very sophisticated objection raised by naturalistic induced theorists. This objection has its origins in the philosophy of mind debate, more specifically in the debate concerning mental causation. Jeagwon Kim and Peter Schulte argue that the position defended here either has to give up the theory of physical causal closure, or it adds another explanatory resource (reasons) in explaining the occurrence of physical events, which then becomes a victim of Ockham's razor—it is a futile explanation because there is already a sufficient explanation that is entirely physical (Kim 2006, 196). Kim calls this the "exclusion"-argument, Schulte the interventionist dilemma. Both arguments rest on the same fallacy: They *identify* the physical events that occur during or at the end of actions with the results of actions and assert that these physical events are the primary *explanandum* of action theory (Schulte 2010, 181). The argument can be reconstructed as follows:

P(1): Physical causal closure: All physical events have sufficient physical causes.

P(2): An action result *is* a physical event E that succeeds a sufficient physical cause C.

P(3): For every m: m is a compelling explanation of E, if it names E's sufficient physical cause C.

P(4): For every p: p is a compelling explanation of E, if it names a person P's *reasons* for doing an action φ (such as murdering Q entailing E).

P(5): Ockham's razor: If there is a sufficient causal explanation m for E, other explanations lack explanatory value.

K: Therefore: p lacks explanatory value.

The first premise is the "physical causal closure"-premise. Kim defines physical causal closure as follows: "If a physical event has a cause (occurring) at time t, it has a sufficient physical cause at t." (Kim 2006, 195) The causal-closure premise is not problematic and it must not be rejected by non-naturalist theories as Schulte suggests (Schulte 2010, 182). The action theory that has been outlined before does not deny that only physical events can cause physical events. We can go even further by saying it is analytically true that only events can cause events. Buildings cannot cause events, universities cannot cause events, but the impact of a bullet can cause an event: It can cause the death of a person. What the non-naturalist theory outlined above denies is that action can be causally explained and it suggests, therefore, that actions require reference to other explanatory resources—reasons. Also, what is not problematic for our non-naturalist theory is P(5), which is Ockham's razor. The problems stem from P(2) and P(3). Consider the picture that underlies Schulte's and Kim's argument:

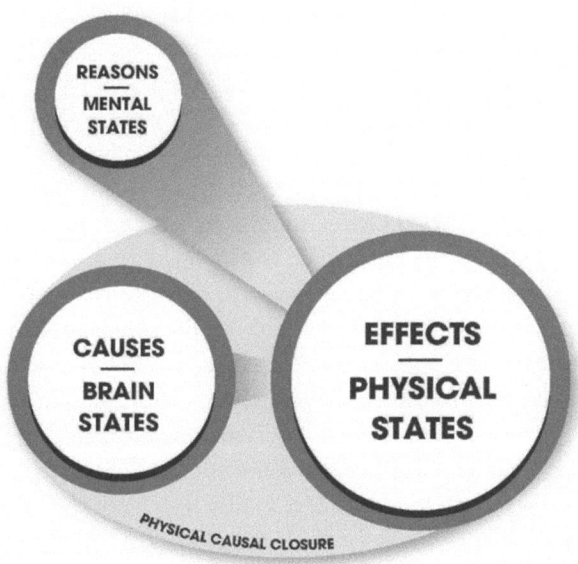

Figure 1: The naturalist view says that causal explanations are sufficient within the causally closed physical "realm." Reference to reasons does not advance these explanations.
Therefore, they are a case for Ockham's razor.

The graph visualizes the situation seen from a naturalist perspective. The "reasons"- approach seems to add something to the explanation of E (the effect or
physical state that occurs at the "end" of the action "Q's death") that is futile,
something that is already entirely explained by denoting their causes, which are
suggested to be physical states—brain states. My critique of Schulte's and Kim's
argument goes as follows: The second premise says that an action's result *is* a
physical event E, and that E has a physical cause C. The problem with this premise
is that it identifies action results with physical events. The naturalist assumes that
the *explanandum* of action theories are physical events, however, this is an *equivocation*. Action results *also entail* physical events, but describing them purely as
physical events means not describing them as action results. Let me clarify this
with the following example: Let us assume that P murdered Q by shooting him.
We could say that the impact of the bullet caused severe loss of blood and severe

loss of blood caused Q's death. This is an explanation of the event "death of Q" by identifying a cause (severe blood loss), and the cause for this event was the impact of the bullet. It is, however, *not* an explanation for the proposition "P murdered Q," it is an explanation for what made Q's heart stopped beating. If saying that severe loss of blood caused Q's death, we explain his death as the effect of a cause and not as an action result. The exact same causal chain ("severe loss of blood" to "Q's death") would be existent, if Q had been fallen into a knife. The causal description of falling into a knife and dying of severe blood loss is the same as in the causal description of being murdered by a gunshot that entails the events "severe blood loss" and "Q's death." Therefore, the explanation of Q's as an action result cannot be substituted by a causal explanation of the presented type. In causal or physical respects, the accidental falling into a knife and the dying from subsequent blood loss is identical to the dying from blood loss having been shot. However, regarding their implications for agency they differ vastly: One is the result of an action—something for which P is highly blameworthy—the other is the result of an accident.

At this point, the naturalist might develop a critique against this kind of reasoning that says: What was presented before as a causal explanation is clearly insufficient but not because it is a causal explanation. The causal chain between "severe blood loss" to "Q's death" is an insufficient action explanation since it only captures *parts of the action*. There could be a more extended description that traces the origins of the action in the agent and particularly the states of his brain. Let me acknowledge this by first refining the description of the exemplary situation. To be more precise from the beginning: the description given exemplarily ("severe loss of blood" to "Q's death") is a causal description but not a physical description of the events. "Severe loss of blood" is not a physically defined concept and neither is "Q's death," or "Q's heart stopped beating." These are concepts of ordinary language. But let us assume, for the sake of argument, that there really is a comprehensive and detailed physical description of these events. Imagine the following situation: You are an inspector and you arrive at a crime scene. You realize a dead person lying on the floor. You ask the forensic investigator about what happened, and she gives you the suggested comprehensive physical description of the events that led from "bullet impact" to "death of Q." Given that you are the inspector, you might become quite angry, because this is clearly not what you wanted to know. What you want to know, in contrast, is whether Q's death has been the result of suicide, murder, or accident. And this question has not been answered by giving the above mentioned comprehensive physical description that underlie the events "bullet impact" until "death of Q." The bullet impact could have been caused by the victim itself, or accidently by playing around with the gun. Clearly, the inspector wants to understand what happened before the gun was fired and the bullet made an impact. Thus far, the forensic investigator provided only a partial description of what has happened. The naturalist could argue that if the forensic investigator had given you a full physical description of *all physical events* from P's decision to murder Q until Q's death, he would have described (and explained) the action "P murdered Q." This description would not leave any questions about whether or not his death was an accident or the result of murder. Such description would be a complete physical description of the action of P's murdering of Q referring to complex causal sequences between physical events.

My objection to this proposal is twofold: First, it must be emphasized that this picture underlies an enormous idealization of the current state of the art of the physical sciences. We have explored the epistemic difficulties of finding causal connections with nomological character in length in the previous sections. Let me recollect some of them: Most of our knowledge of our body is biological and chemical knowledge. There are several theories of the working mechanism of muscles. It is suggested, for instance, that chemical energy is translated to mechanical energy in the actin-myosin-complex of the muscle. Such theories also exist about the working mechanisms of neurons. Neurons are suggested to stimulate other neurons with electrical charges by building up sodium-potassium potentials.[47] Such knowledge about the bio-chemical working mechanisms is highly diverse and fragmented. It is expressed primarily in biological and chemical concepts and is gathered through a variety of different research methods and experimental approaches (MRIs are just the most famous). Even if there were a comprehensive theory of bodily motion in physical terms—which presumes that there is

---

47   Note that a lot of this research employs intentional idioms in its language like in
       this exemplary statement: "Neurons stimulate each other with electrical charges."
       (Janich 2009, 73)

essentially an adequate translation from these diverse bio-chemical theories about muscle contraction and neurons to physical theories despite their different ways of experimenting and measuring—it would still be a long way to go for translating such knowledge into a *causal theory*. Remember that most of our physical knowledge is expressed in "laws" that relate *universals* like velocity and time. The best way to interpret such "laws" is to regard them as instruments to calculate and predict the occurrence of events in ideal, shielded contexts (Chalmers 1990, 66). Thus, if we take, for example, the expression "muscles translate chemical energy into mechanical energy in their actin-myosin-complex" there is still a long way to go to provide a *causal description* of muscular movement. Remember that causal theories relate *events* to each other and not abstract objects like "chemical energy" to other abstract objects like "mechanical energy." Ambitious naturalists might claim that both the reduction of these bio-chemical descriptions to purely physical descriptions about, for instance, molecular movement and the expression of such knowledge in causal terms are possible. The naturalist might claim that there can be such a description of causal chains leading from the P's decision through his brain and body to the pulling of the trigger, the exiting of the bullet to its impact, the loss of blood, and the death of Q.

Let me challenge this optimistic view with another illustrative example where both the beginning and the end of the action are even less clear to be identified in physical terms. Take the following action: "Deliberating the philosophical mind-body problem." Many philosophers permanently perform this action. They spend years or lifetimes deliberating about the mind-body problem. The idea that there could be something like a complete physical description of this action leading from the initial thought to its end seems entirely miraculous. When does such an action end physically? Is it finished with the writing of a book or the publication of an article about the mind-body problem? When does such an action even start? These hindrances should be taken seriously. They are not merely of a technical, but of a more fundamental nature. The obstacles of describing the abundance of physical events that occur during an action of the kind "deliberating the philosophical mind-body problem" are significant. Naturalists tend to overestimate the success of the natural sciences when assuming that we just have to think of explaining actions analogously to, for example, the way we explain the changes of consistency of substances like water when cooled down or heated (Searle 1991, 15).

But as mentioned before, many of these explanatory successes are limited by their applicability only to enclosed experimental situations which can hardly, if ever (remember the example of deliberating the mind-body problem), be construed to investigate human *action* (Chalmers 1990; Janich 2009, 160). A naturalist might still say that we have these types of explanations, and nothing speaks *in principle* against having such explanations for actions too. However,—and this is my second objection to the naturalist strategy—to argue that such physical explanation can exist *in principle* equals saying that actions can be causally explained. Moreover, this assertion begs the question of agency. It does not show that actions can be causally explained; this stance presupposes it. In contrast: Even enormously detailed and complex physical descriptions of the events that happen during an action would not be descriptions of the action, of the willed doing of something of a person.

My theory can be upheld without disposing the principle of physical causal closure. It relies on the assumption that physical events are not the only existing entities in the world. There are actions and agents too. Our causal knowledge is so fragmented that it offers plenty of space for agency, and there are no reasons to assume that the world consists entirely of sets of causes and their effects (which would be the determinist view). To visualize this approach, and to put it into stark contrast with the picture suggested by the naturalist theorists, let me introduce the following graph:

Figure 2: The non-naturalist view suggests that reasons provide sufficient explanations for action results. Physical events that occur simultaneously to action results can be approached with distinct causal explanations.

The upshot of this view is that there are causal connections and causal explanations of some of the *sub-events* that occur during actions: The bullet impact causes blood loss and blood loss causes death. This is an ordinary causal explanation that can be true without assuming that it is an explanation of the action—whatever the action was. Similar to this, there are numerous other causal events happening in our brains and our body during action. Providing a detailed and complex causal description of all these events would still not amount to a description of the action. Causal chains of nomological and non-nomological type run as visualized in the previous graph through our bodies and around us while we act (Keil 2000, 457). These causal chains do not normally interfere with our doing of things (although they occasionally do). In contrast, we often rely on them: When P murders Q by shooting him, he presumes that the impact of the bullet will be fatal as is usually the case. My theory relies essentially on a parallelism of agency and causation, and I do not believe that there are good reasons to reject such a view without pre-suming naturalism beforehand. The explanatory resources to analyze actions are the reasons of the doers. The fact that we come to our decision through reasoning is so familiar to us that we sometimes forget about this as a proper explanatory resource. The success of the natural sciences in reproducing regularities in exper-iments and technological devices probably contributed to the urge to press actions into a causal scheme. However, this is not the only explanatory scheme to make the world intelligible. To put it a bit bluntly: The role of causal explanations in the entirety of our scientific knowledge is overrated. Even the natural sciences make use of other explanatory resources and many of those ordinary candidates for causal laws, such as the law of steady motion, are not instances of causality. Nat-uralism broadly understood also employs non-causal explanations like narrative (historical) explanations which are familiar in biology (Toepfer 2013, 34). Fur-thermore, in non-scientific contexts we often rely on chance when explaining cer-tain occurrences (Hampe 2007, 24–26). Even chance is sometimes employed as a resource to make the world intelligible. Explanations of actions with reference to chance would undermine the authorship condition of agency as much as causal explanations. It should be exemplified with reference to the narrative explanations in biology and explanations by chance that there are a varieties of ways to make the world intelligible that are non-causal explanations. The previous account simply requires that we accept justifications of actions through reasons as expla-nations *sui generis* in addition to the already established explanatory variety. The fact that the natural order contains agents and actions is a good reason to believe that determinism is not a comprehensive theory of the world.

4.5   Varieties of Compatibilism

Roughly speaking, compatibilism is the belief that determinism and moral respon-sibility can be reconciled. In the following, I will first outline the baseline of com-patibilist's reasoning and then discuss three approaches in more detail: the first is John Martin Fischer's "significant distinction"-argument, the second is Moritz Schlick's theory of "effect compatibilism," and the third—which will be discussed in chapter four together with Thomas Nagel's reasoning—is the "theory of the moral sentiments" of Peter Strawson. In general, compatibilists believe that moral responsibility does not require genuine free will. Moral responsibility can rely entirely on the inextricability of our moral sentiments (Strawson 1962), the differences between "normal" and in a sense "impaired" people, which prevail even if determinism were true, or an understanding of praising and blaming as the social demand for reformation of the agent and his behavior, as reasons of deterrence and protection (Schlick 1962c; Fischer 2007). According to compatibilist reasoning, those aspects of our evaluative moral system can be upheld (or are inescapable), which makes the possible truth of determinism irrelevant. Many compatibilists share the common belief that one cannot make sense of the idea of genuine free will. Compatibilists claim that the opposition of the principle of causation is chance, and if actions were the results of chance, this would equally undermine freedom (Schlick 1962c, 156; Tugendhat 2007, 62). Hume writes in this manner:

> Had not objects a regular conjunction with each other, we should never have entertained any notion of cause and effect; and this regular conjunction produces that inference of the understanding, which is the only connexion, that we can have any comprehension of. Whoever attempts a definition of cause, exclusive of these circumstances, will be obliged either to employ unintelligible terms or such as are synonymous to the term which he endeavors to define. And if the definition above mentioned be admitted; liberty, when opposed to necessity, not to constraint, is the same thing with chance; which is universally allowed to have no existence. (Hume 1975, 96, 2003, 123)

In a chapter on responsibility in his *Ethics: The Invention of Right and Wrong*, Mackie depicts the same thought:

> Besides, if strict determinism is not true, the most likely alternative is a partial determinism mitigated by a certain amount of randomness - an Epicurean physics. But this would be equally incompatible with the notion of ultimate responsibility. Such responsibility would evaporate if we tried to attach it to purely random occurrences just as clearly as it would disappear to infinity along causal chains. It requires for its resting point a contra-causally free and yet determinate and active self, the concept of which it is so hard to render coherent. Equally, the notion of an absolutely open choice is as incompatible with an Epicurean physics as with strict determinism: the metaphysical self would be left idle and imprisoned by chance no less than by law. (Mackie 1990, 226)

Mackie also suggests that there is incoherence in the idea of ultimate responsibility, which is often conceived by compatibilists as a "contra-causal" power of the agent, who can "miraculously," as Hume says, suspend the laws of nature (Hume 1975, 114; Ayer 2013, 318). However, as a compatibilist Mackie claims that our beliefs in moral responsibility have a solid basis that cannot easily be undermined. He denies that determinism's truth would make a difference for moral responsibility:

> In particular the distinction between intentional and non-intentional action, and all that gives moral significance to this distinction, could still stand if determinism were true. (Mackie 1990, 226)

What the origins of these intentions are, however, Mackie leaves unanswered. As argued before, an appropriate theory of action must understand these intentions as the results of the deliberations *undertaken by* the agent; otherwise, they are not his intentions, and their deliberation is merely an illusion. Although, the term illusion does not quite capture the problem: The active dimension of agency is just not covered by a deterministic viewpoint. Mackie does not help us find a notion of intentionality that does not require freedom. Instead, he maintains the distinction between intentional and non-intentional actions can be upheld and leaves us uninformed about their sources, which were *ex hypothesi* identical in a determined world: preceding causes. In a certain sense, however, the distinction between intentional and non-intentional action can be clearly upheld. We would still be able to say that someone had the intention G and did φ accordingly, if determinism were true. Yet, in which sense could we still speak of these intentions as the intentions of that person? We previously argued that in order to be morally responsible those intentions must be neither the result of external forces nor a result of a lack of alternatives nor of an impairment of one's capacities to reason. But if determinism were true, these capacities were not the capacities *of* the agent. The agent might still have the "feeling" of being the author of his decisions and actions, yet the results of his deliberations and reasons are predetermined, and even if she believes that *she* has arrived at the proper practical conclusions, they have not been hers to arrive at in an ultimate sense. Mackie is certainly right that there is (and will remain, if determinism turns out to be true) a difference between something that happens entailing such type of reflection or deliberation, and an accident or process similarly devoid of intentionality. A "normal" person who undergoes a process of deliberation and "acts according to it" is not detached from his actions in the same way in which, for instance, an alcoholic can be detached from his behavior. The alcoholic might deliberate about his misery, pity himself and deeply desire to live a fundamentally different life and still, after hours of deliberation, take the bottle of alcohol and get drunk.[48] He will rightly say afterwards: "I could not help it. The urge was overwhelming." This clearly does not resemble the situation in which a normal person would find herself in when acting, if determinism were true. "Normal" people under "normal" circumstances will not *feel* coerced, manipulated or pushed; they will not become spectators of their own behavior, which is allegedly how many addicted people experience their addiction (Bieri 2013, 99). For a compatibilist, this *experience* of being the originator of one's actions is allegedly the only "freedom worth wanting." (Dennett 1990; Kane 1996, 15) Regarding this phenomenon of subjectively experienced freedom, John Martin Fischer says that the truth of determinism would not "necessarily manifest itself to me phenomenologically." (Fischer 2007, 46) It is in this regard also clear that social sanctions are in a different sense "compelling" than the laws of nature would be, if they applied universally. The laws of nature do not "prescribe" as Schlick rightly states (Schlick 1962c, 147). Thus, many compatibilists, such as Hume, Schlick, and Ayer oppose freedom not to what they call the causal principle (which is already not quite the same as determinism) but to external compulsion and coercion (Schlick 1962c, 148–150; Hume 1975, 95; Ayer 2013, 319). Hume writes nonchalant about the thoughts of a prisoner, who is sentenced to death and deliberates his options:

> Here is a connected chain of natural causes and voluntary actions: but the [prisoner's] mind feels no difference between them in passing from one link to another [when deliberating]: Nor is less certain of the future event than if it were connected with the objects present to the memory or senses, by a train of causes,

---

48   One does not have to be an alcoholic or kleptomaniac to understand this phenomenon as a lack of "control." Susan Wolf describes it as follows: "Kleptomaniacs and victims of hypnosis exemplify individuals whose selves are alienated from their actions [...]." (Wolf 2013, 332)

cemented together by what we are pleased to call a *physical* necessity. (Hume 1975, 90)

Then again, any attempt to explain the difference between a "normal" person, for example the prisoner, and an alcoholic or a kleptomaniac will necessarily utilize notions such as "controlling one's behavior according to foregoing deliberation" or "acting according to foregoing balancing of reasons." (Wolf 1990, 10) When we make use of such notions of control, we cannot consistently consider these capacities as the capacities *of that person*, if determinism were true, they would not originate from her and, therefore, speaking of the "actions of people," in a determined world is already suggesting a terminological confusion (Keil 2007, 79).[49] A worldview that suggests being comprehensive and describes only processes cannot account for the existence of agency, as mentioned before. Yet, John Martin Fischer believes that compatibilists can clinch to this important distinction, if determinism turns out to be true:

> A compatibilist can maintain this distinction, even if it turns out that the physicists convince us that the probabilities associated with the relevant conditionals – the conditionals linking the past and laws with the present in physics – are 100 percent, rather than 99.9 percent. And this is a significant and attractive feature of compatibilism. Incompatibilism would seem to a collapse of the important distinction between [normal] agents such as Sam and thoroughly manipulated or brainwashed or coerced agents. A compatibilist need not deny what seems so obvious, even if the conditionals have attached to them 100 percent: there is an important difference between agents such as Sam, who act freely and can be held morally responsible, and individuals who are completely or partially exempt from

---

49   Keil writes that our ordinary notion of action already has massive implications regarding the metaphysics of freedom. He argues that there is not a separate problem of freedom, if it is true that we can act (Keil 2000, 329, 2007, 89). The common understanding of "P has done φ" already implies a genuine ability of "could have acted otherwise."

moral responsibility in virtue of *special* hindrances and disabilities that impair their functioning. (Fischer 2007, 48)[50]

Whether it is meaningful to assume that the truth of determinism can come in degrees or probabilities should not be discussed here. What can be said about this position in light of the reasoning presented in the previous sections is that there are certain events, which are more loosely connected to their causes than other regularities. The plausibility of determinism, however, depends on the plausibility of the existence of the natural laws, and this did not seem to be a creditable idea. However, the more pressing question here is, as before: In virtue of which facts can the distinction between the normal agent and other "impaired" agents be upheld? If the resilience of this distinction is such an attractive feature of compatibilism, then compatibilists have to explain wherein the difference between agents that are morally responsible and those that are not lie in, and why they make a difference, *even if* determinism were true. What libertarians demand at this point is a thorough analysis of the meaning of being a "normal" agent and, thus, a morally responsible one. Fischer sets out the following example of Sam, who "grew

---

50   Strawson's extended version of this argument in Freedom and Resentment is sometimes overlooked, for example by Geert Keil. Strawson writes: "Now it is certainly true that in the case of the abnormal, though not in the case of the normal, our adoption of the objective attitude is a consequence of our viewing the agent as incapacitated in some or all respects for ordinary inter-personal relationships. He is thus incapacitated, perhaps, by the fact that his picture of reality is pure fantasy, that he does not, in a sense, live in the real world at all; or by the fact that his behaviour is, in part, an unrealistic acting out of unconscious purposes; or by the fact that he is an idiot, or a moral idiot. But there is something else which, because this is true, is equally certainly not true. And that is that there is a sense of 'determined' such that (1) if determinism is true, all behaviour is determined in this sense, and (2) determinism might be true, i.e. it is not inconsistent with the facts as we know them to suppose that all behaviour might be determined in this sense, and (3) our adoption of the objective attitude towards the abnormal is the result of a prior embracing of the belief that the behaviour, or the relevant stretch of behaviour, of the human being in question is determined in this sense. Neither in the case of the normal, then, nor in the case of the abnormal is it true that, when we adopt an objective attitude, we do so because we hold such a belief." (Strawson 1962, 13)

up in favorable circumstances," and "has no unusual neurophysiological or psychological anomalies or disorders," and is not "brainwashed, coerced, or otherwise 'compelled' to do what he does." (Fischer 2007, 47) Fischer continues to argue that Sam "deliberates in the 'normal way' about whether to deliberately withhold pertinent information on his income tax forms, and, although he knows it is morally wrong, he decides to withhold the information and cheat on his taxes anyway." Fischer concludes on the same page of his article, "[i]nsofar as Sam selected his own path, he acted freely and can be held both morally and legally accountable for cheating on his taxes." (Fischer 2007, 47) The libertarian will astonishingly react to this proposal and ask: "If determinism were true, why exactly should we believe that Sam has selected his own path and why should we believe that *he* decided to withhold information?" If determinism is true in the way the libertarian claims, then whatever Fischer attributes to Sam as allegedly his actions were not up to him.

In a sense, these underlying determined processes might have left the *perception* of him as the agent of his actions unaffected. From the agent's internal point of view the future will still look open, and it will still "feel" as if possible alternatives in this future are realized through his decisions and actions (Nagel 1986, 114–115). But if determinism were true, the results of his deliberations are fixated long before his experience of deliberating about tax evasion actually begins and *this* starkly conflicts with our understanding of UR and AP (Mele 1999, 285). Imagine again an epistemicly perfect being; such being would know a thousand years in advance what Sam will do (or better: what will happen within and through him) and which reasons will "move" him to disclose tax information. From this point of view, his deliberations were neither open-ended, nor did *he* play any role in determining the outcome. The fact that determinism does not undermine the phenomenological experience of authorship does not increase our confidence in compatibilist responsibility, as much as it does not become less pitiful when a person who is drugged repeatedly assures that he is the president of the United States and loves his job, if in reality he is anything but that. A strong illusion remains an illusion nevertheless (Seebaß 1993a, 244). Let us discuss a different but also common line of compatibilist thought that strongly utilizes the concepts of praise and blame as means of social regulation. This compatibilist theory could be called—

following Saul Smilansky's terminology "effect compatibilism," or in Susan Wolf terms "pragmatic compatibilism." (Wolf 1990, 20; Smilansky 2000, 27)

4.6   Effect Compatibilism and Psychological Determinism

The theory to be discussed in the following has most recently been advanced under the label "revisionism." Manuel Vargas, for instance, summarizes the basic aspects of what is in the following considered as "effect compatibilism" in the so-called *responsibility system* as follows:

> [...] the responsibility system aims to get us creatures like us to better attend to what moral considerations there are and to appropriately govern our conduct in light of what moral reasons those considerations generate. Assessments of praise-worthiness and blameworthiness are not merely reactions we happen to have to one another [...]. They play a special role in getting us to be better beings, agents better attuned and more appropriately responsive to moral considerations and the reasons they generate. (Vargas 2007, 155)

In these basic convictions, Vargas' revisionism resembles the viewpoint of the authors that should be discussed in the following. Revisionism claims that the responsibility system contains reciprocal practices that aim at improving human behavior in moral respects that are reasonable without libertarian commitments (ibid., p. 156), and this is also a central conviction of effect compatibilism. In the present section I will focus on more traditional accounts of effect compatibilism, namely those of Moritz Schlick and David Hume because it seems that their theory is often misunderstood and too readily associated with forms of utilitarianism. This is an injustice to their theory. For a variety of other reasons, which should be outlined in this and the next section, however, I do not believe that effect compatibilists can persuade a libertarian thinker to let go of his idea of genuine freedom and these reasons also apply to Vargas' revisionism. Revitalizing and defending Schlick's and Hume's theories against common charges will help me to clarify their viewpoints and also to sharpen my own.

Effect compatibilist emphasize the role of moral and legal sanctions *as means* of reformation and deterrence, and at first sight it seems as if they plainly equalize being responsible with being the appropriate object of punishment and blame. This interpretation is an exaggeration, since many effect compatibilist merely take the sanctioning role of blame and punishment as a descriptive fact about our societies, while at the same time providing necessary conditions for appropriately believing in people's responsibility. Before turning to the "effect dimension" of the theory and exploring the relation between *being* responsible and being *held* responsible in more detail, we should assess the proposed conditions and explore the fundaments of the theory, which is the so-called idea of "psychological determinism." Some effect compatibilists such as Moritz Schlick believe that the debate about freedom of will constitutes a pseudo-problem:

> [...] this pseudo-problem has long been settled by the efforts of certain sensible persons; [...] with exceptional clarity by Hume. Hence it is really one of the greatest scandals of philosophy that again and again so much paper and printer's ink is devoted to this matter [...]. Thus I should truly be ashamed to write a chapter on "freedom." (Schlick 1962c, 143)

Schlick makes here explicit that his approach owes most to Hume's philosophy. However, Hume relies much more on the idea of the moral sentiments, as I will recall in more detal in section 5.4. In addition, Schlick utilizes the notion of our *"feeling* of having been able to act otherwise." This feeling does not primarily address the behavior of *other people* as objects as Hume's moral sentiments do, but concerns merely the relation to one's *own* actions and *consciousness* of their voluntariness (Schlick 1962c, 129). Both Hume and Schlick emphasize the function of the practices of praising, blaming and punishing in our societies. They share the idea that human action can only be explained with reference to the regular conjunction of motives or desires and action. Schlick calls these regularities "psychological laws." (Schlick 1962c, 144) Both authors claim that these conjunctions are necessary connections between motives or desires, and human actions, while in both theories the concepts "motive" and "desire" are relatively vague and interpreted differently. Hume, who—in contrast to Schlick—repeatedly expresses a certain skepticism about the nature of causality—he was primarily interested in understanding how we acquire the habit of induction (Hume 1975, 73–75; Strawson 1985; Keil 2007, 53))—and the idea of the "necessary conjunction," hardly ever speaks of laws as Schlick does. Both Hume and Schlick claim, however, that "psychological determinism" is required for any science that aims at understanding of human behavior and for morality, which both suggest to be the science that aims primarily at *influencing* human behavior (Schlick 1962c, 156–157). Hume writes:

> Where would be the foundations of *morals*, if particular characters had not certain or determinate power to produce particular sentiments, and if these sentiments had no constant operation on actions? And with what pretence could we employ our *criticism* upon any poet or polite author, if we could not pronounce the conduct and sentiments of his actors either natural or unnatural to such characters, and in such circumstances? It seems almost impossible, therefore, to engage either in science or action of any kind without acknowledging the doctrine of necessity, and this *inference* from motives to voluntary actions, from characters to conduct. (Hume 1975, 90)

If those regularities would not hold, chaos would prevail. Schlick says: "If decisions were causeless there would be no sense in trying to influence men; [...]." (Schlick 1962c, 157) In his book *Enquiries Concerning Human Understanding* Hume emphasizes several times the importance of psychological regularities for punishment and responsibility (see also his *Treatise on Human Nature* (Hume 2003, 325)):

> All laws being founded on rewards and punishments, it is supposed as a fundamental principle, that these motives have a regular and uniform influence on the mind, and both produce the good and prevent the veil actions. We may give to this influence what name we please; but, as it is usually conjoined with the action, it must be esteemed a *cause*, and be looked upon as an instance of that necessity, which we would her establish. (Hume 1975, 97–98)

In this theory, punishment and reward are conceptualized and reinterpreted as motives that *cause* people to behave in certain ways. Hume's concept of a "uniform" or "regular influence"—the very widely applied principle of causality—expressed here is far from the previously discussed idea of determinism and clearly distinct from a realist interpretation of the laws of nature. Schlick on the other hand, believes that "[t]he natural law is [...] a description of how something does in fact behave." (Schlick 1962c, 147) In contrast, Hume promotes a nominalist concept of causality which is a much weaker notion, as mentioned before (Keil 2000, 157–

160; Hampe 2007, 75). In any case, it is grossly misleading to interpret external sanctions, such as punishments and rewards as *determinants* of actions as Hume and Schlick do. People's capacity to take possible sanctions as *a reason* for or against action into consideration can also be interpreted as a fact about people's general rationality. This matter of fact just appears from an external point of view to be very similar to typical causal propositions about other regularly conjoined events, such as the changing of seasons. It is true that most people are not systematically irrational; they expose consistency in their behavior, they follow more or less strict principles (principles of prudence and of morality) and do not change their habits on a daily basis, act erratic, or expose permanent impulsive behavior. These aspects might even be constitutive conditions for personhood (Williams 1981c; Bieri 2013, 65). Thus, people go regularly to work, they eat meals with fork and knife (in Europe) and they order pizza when they are hungry. In this sense, their behavior is *regular* and *predictable*. However, these simple facts about regularities in human behavior should not to be understood as determining factors (Campbell 1967a, 46). Not only, can people omit or change these behavioral patterns to a significant degree which is sometimes encouraged, but, moreover, many actions (in contrast to mere behavior) can only be understood as willed decisions that are made *against* initial desires and inclinations (Campbell 1967b, 74; Tugendhat 2007, 62). Thus, a libertarian is not forced to deny that there are regularities in human behavior, as long as she emphasizes the difference between this thesis and the idea that the outcomes of *deliberate decision-making* are not predetermined.[51] The thesis of psychological determinism passes, in this sense, the actual object of concern of the freedom of will debate, when focusing on *behavioral* patterns. The object of concern of the free will debate are reflected decisions often made against initial inclinations or established dispositions, which can then indirectly shape those behavioral patterns. For example, one can eat meat for decades without ever questioning this practice and still one day realize (or being made aware by friends to realize) that there is something morally wrong with eating

---

51   Keil writes in agreeable manner that for rationalizing human behavior, we do not have to oppose strict regularities with chaos, as Hume does, but with soft variability (Keil 2000, 350).

meat, get informed about alternative diets, and find ways to steadily replace animal products from ones menu until one has established a disposition for a vegetarian or vegan diet. We could call this—following Kane—a self-forming action (Kane 2007, 14). Clearly, it is *then* predictable what that person is going to order in a restaurant—a vegetarian dish. In a certain sense, it would also be true, if that person admits that she *can no longer* eat meat. This, however, does not imply that it has been predetermined for this person to become a vegetarian. She is responsible for having consciously made this decision, even if will not always reflect on it thereafter. We would not be willing to say that this person has decided to be a vegetarian, if she does not start to order meat regularly from then on. If this person does not start ordering meatless dishes, we are inclined to argue that she has misunderstood the concept of vegetarianism or not entirely absorbed what it means to be a vegetarian. Given that the theory of psychological determinism claims to provide a descriptive law that conjoins motives or desires and actions one might—given this description—get the impression that these "psychological laws" can be endlessly supported by *ceteris paribus* conditions like "someone has to *entirely* absorb a desire" or "the person ought to have the appropriate beliefs to a desire," to rescue it against counterexamples where the motive-action pairs divert from the suggested regularity.

To understand this objection, consider the possible responses of a psychological determinist to a situation in which a vegetarian person diverts from her meat avoiding behavior. It could be that his desire was never intended to be a *strict* vegetarian, or that this person has realized the existence of overriding reasons in this particular situation, which she also desires to respect. This makes any particular psychological regularity indivisible against all kinds of counterexamples: In short, psychological determinism is, from this perspective, *not falsifiable* (Hartmann 2005, 23). Considering a number of reasons, it should be clear by now that the standard argument for psychological determinism—the regular conjunction of motives and actions—is not conclusive for regarding motives as determinants of actions (Chalmers 2007, 172). Furthermore, during the vegetarian's deliberation, they might have compared the monetary costs of each diet set by the industry and policy regulations and taken this comparison into account. However, this does not mean that the monetary costs have determined their decision either. Such external factors are becoming the reasons she balances throughout her deliberation. It should be clear that the majority of things people do are not deeply reflected and permanently alternated (people do not change their diet every day) and this makes human behavior largely regular and predictable. But as argued before, this misses the target of concern of the free will debate. Free will is at stake when ends are deliberated and intentions formed, and they can be directed at a person's attitude and, for instance, their eating habits or other dispositions. Psychological determinists, such as Schlick and Hume, must show that the *actions* that form those behavioral patterns, such as the decision to become vegetarian, are predetermined, and that the external factors that appeared in my example as reasons for the agent are in fact determinants in a much stronger sense than ordinarily suggested (Keil 2007, 45).

Although Schlick and Hume often articulate this point in an ambivalent manner, they are both unable to make sense of the idea that reasons can be motivating factors of human action.[52] How is the theory of psychological determinism related to moral responsibility in Schlick's argumentation? Schlick argues in a similar manner as Hume that holding someone responsible aims at influencing this person's conduct by modifying his motives:

> Punishment is concerned only with the institution of causes, of motives of conduct, and this alone is its meaning. Punishment is an educative measure, and as such is a means to the formation of motives, which are in part **to prevent the wrongdoer from repeating the act (reformation) and in part to prevent others from committing a similar act (intimidation)**. [...] Hence the question regarding responsibility is the question: Who, in a given case, is to be punished? Who is to be considered the true wrongdoer? [...] the "doer" is the one *upon whom the motive must have acted* in order, with certainty, to **have prevented the act**

---

52   Schlick writes: "Of course the processes whereby the general welfare becomes a pleasant goal are complicated, and one must not, above all, attribute too great a role to rational insight. For even if men thought much more accurately than they usually do about the consequences of action, such considerations would have but little influence in the realm of feelings [own emphasis]." (Schlick 1962b, 98) The crucial question is then: Is there little or no influence? Hume writes in his Treatise that reasons cannot be practical, but only "guide" our judgments: "Morals excite passions, and produce or prevent actions. Reason of itself is utterly impotent in this particular. The rules of morality, therefore, are not conclusions of our reason." (Hume 2003, 325)

> [...].The question of who is responsible is the question concerning the *correct point of application of the motive* [own emphasis]. (Schlick 1962c, 152–153)

A *narrow* interpretation of this passage zeros in on the aspect of preventing the doer and others from executing similar acts. By focusing on this aspect of Schlick's compatibilist theory of responsibility, the entitlement of effect compatibilism appears to be suitable and comes to the forefront. *Prima facie*, however, this move suggests some absurd implications. Regarding this narrow interpretation; Schlick's theory seems to blur the boundaries between educating, coercing, stimulating, nudging, domesticating and holding responsible. Thus, one might think of the myth about the Persian king Xerxes narrated in the histories of Herodotus (Herodotus 1958, 20). To conquer and defeat the Greek, Xerxes wanted to lead his army over the Hellespont to Sestos. Thereto, he ordered to build a bridge over the gulf. Directly after the construction was finished, a storm pushed the water from the gulf with an immense power against the bridge and destroyed it. Herodotus tells us that Xerxes was so angry about the setback that he ordered to behead the two master builders of the bridge. Furthermore, he commanded to whip the water at the Hellespont for this immense iniquity. Three hundred soldiers were told to stand in the water and whip it. Herodotus does not tell about the initial reasons of Xerxes' order of this penalty. It can be suggested that he just wanted to release his anger about the delay of his victory. In any case, Xerxes reaction seems to be plainly absurd. Any theory that takes the concept of "being held responsible" so far off our common usage cannot be acceptable. However, one might say—in the spirit of the previous argument—this is only because there is no way of *affecting* the sea's behavior by whipping the water. Then, one could argue that building a dike equals holding the sea responsible: It "influences" its behavior.

Schlick does not have to comply with this excessive application of the term responsibility for several reasons. He says (extremely ambivalent though) that only the doer of an action "upon whom the motive must have acted" is an appropriate object of such reaction, and this certainly does not apply to inanimate objects. But if we presume a wide understanding of the concept "motive," we might find that animals or children also "act" under the regular influence of their motives: They eat when hungry and drink when thirsty, and they are also "able" of changing their behavior, whenever their motives change (Wolf 1990, 7).[53] This objection is fueled because Schlick seems to equate different concepts, such as "motive," "desire," and "natural tendencies," throughout his arguments. Thus, Charles Campbell focuses on this aspect and objects to Schlick:

> We do not ordinarily consider the lower animals to be morally responsible. But *ought* we not to do so if Schlick is right about what we mean by moral responsibility? It is quite possible, by punishing the dog who absconds with the succulent chops designed for its master's luncheon, favourably to influence its motives in respect of its future behaviour in like circumstances. If moral responsibility is to be linked with punishment as Schlick links it, and punishment conceived as a form of education, we should surely hold the dog morally responsible? The plain fact, of course, is that we don't. We don't, because we suppose that the dog 'couldn't help it': that its action (unlike what we usually believe to be true of human beings) was simply a link in a continuous chain of causes and effects. (Campbell 1951, 447)[54]

Against the accusation of distorting the distinction between holding morally responsible and domesticating or educating, Hume argues that our moral sentiments rebel against the idea of holding animals and—as he explains with reference to the example of a tree that overtops and destroys its parent—other inanimate objects responsible (Hume 1975, 293; Russell 1990, 547). I will discuss this later on in length on conjecture with Peter Strawson's theory of the moral sentiments. If

---

53   Note, that Manuel Vargas explicitly encourages such transfer when writing that dogs, which are usually considered as being "determined," can by habituation learn appropriate behavior. The cognitive differences between humans and dogs "do not provide a reason for thinking that sensitivity to considerations and the ability to appropriately govern one's behavior in light of them could only be had by humans in an indeterministic universe." (Vargas 2007, 157) He does not seem to realize that we attribute meaning to the practice of learning dog tricks without believing in their responsibility. In Vargas' revisionist account, which is based on a denial of libertarian freedom (ibid., p. 143), one cannot even make sense of the distinction proposed below between believing in someone's responsible and what he calls the "responsibility system." The instances of believing in someone's responsibility and holding him responsible are identical in this theory. I think that my reasoning below makes intelligible, why there is indeed a distinction between them.

54   See also (Taylor 2013, 309).

Schlick accepts the wide application of the concept "motive" to also cover the motivational urges of animals and children, he can still refer to other important aspects of his theory of responsibility that might protect him from the allegation of entertaining arguments that lead to a *reductio* of this kind: One of them is the phenomenological side of agency—the feeling of having acted freely—, the other one is the related feeling of having been able to act otherwise, which he endorses both. These aspects are not adequately acknowledged in a narrow understanding of his theory that zeros in on the sanctioning dimension of effect compatibilism. The consciousness of responsibility, the feeling of having acted freely—which is according to Schlick even more important than the sanctioning dimension—is emphasized in more detail later on in the text:

> But **much more important** than the question of when a man is said to be responsible is that of when he himself **feels** responsible. Our whole treatment would be untenable if it gave no explanation to this. [...] What is this **consciousness of having been the true doer of the act**, the actual instigator? Evidently not merely that it was he who took the steps required for its performance; but there must be added the awareness that he did it "independently," "of his own initiative," or however it be expressed. This feeling is simply the consciousness of *freedom*, which is merely **the knowledge of having acted of one's *own* desires**. And "one's own desires" are those which have their origin in the regularity of one's character in the given situation, and are not imposed by an external power, as explained above [when being threatened]. The absence of the external power expresses itself in the well-known **feeling** (usually considered characteristic of the consciousness of freedom) **that one could have acted otherwise** [own emphasis]. (Schlick 1962c, 154–155)

If knowledge is understood here as propositional knowledge, then animals probably cannot entertain knowledge of their own freedom, although they might "feel" as uncoerced as normal human beings might. An alcoholic will argue that his urge to drink overwhelmed his reasonable desire not to do so. He will claim that he did not "feel" as if he acted on his own will. Schlick argues in another paragraph that alcoholics, people who are influenced by drugs or someone who is mentally ill are influenced by factors that "hinder the normal functioning of [their] natural tendencies." (Schlick 1962c, 151) Even though such people sometimes feel (when they are drugged) that they act according to their own desires, they are impaired in their "normal functioning." However, it is far from clear what is meant with the "normal functioning of the natural tendencies." Schlick is utterly imprecise in his usage of this concept throughout his text. The problem of vagueness of the terms "motive," "desire," and "natural tendencies" that build the theoretical basis of his theory of action on which, subsequently, his notion of responsibility becomes obvious again at this point. One might claim that the dog plainly follows his "natural tendencies," feels probably uncoerced and can be influenced through reward and punishment. He fulfills the conditions for the weaker concept of "freedom of conduct." Still, we do not think that dogs can help doing what they do. To resolve this, Schlick would have to employ a stronger notion of agency, which he rejects as being impossible. Before, I developed a brief critique on Schlick's theory. Let us first resemble the entailments of Schlick's theory of psychological determinism and contrast those aspects critically with the view of free action that presented in previous sections. For Schlick, the agent is the bearer of desires and the desires "determine" what the agent does. The agent feels responsible and uncoerced, which is most important to be responsible. Furthermore, the agent also feels that he could have acted differently, which is similarly important for being responsible.

This reasoning cannot obscure the fact that in Schlick's and Hume's theory of agency as presented above, the agent is not a doer but merely the locus of his own motives that "act upon him." If we want to make sense of action, we must conceive the agent as being able to balance reasons, to arrive at its own intentions, and act accordingly to them. In short, we need the freedom to choose what to intend and what to do. The conception of the agent as the locus of motives that overcome it, such as a coughing fit or a headache, allegedly forego its behavior, which undermines his freedom beforehand. Agency must be conceived as something *active* and not as something passive (Hampe 2007, 176). Thus, Schlick cannot consistently say, as many other compatibilists do, P could have done something different, but only something different could have *happened* through P's alternated motives (Pauen 2004, 129). Moreover, as I argued before, our feeling of acting freely—what Schlick calls the "awareness of doing something on one's own initiative," which is such an important pillar of the argumentation in the above cited passage—cannot mascaraed the fact that determinism would eventually disclose the possibilities that occur as alternatives during deliberation (Seebaß 1993a, 244). In this respect, Schlick's theory can be seen as a predecessor of Fischer's "phenomenology of agency"-approach outlined before. The similarities are striking.

Still, it must be repeated that there is only one future, if determinism were true despite the introspective feeling of the open-endedness of the future and the *feeling* of authorship.

Without being able to discuss the following issue in detail within the scope of this enquiry, it should be mentioned that other compatibilists entertain a notion of "could have done otherwise" that does not rely—as in Schlick's and Fischer's case—on one's *feeling* or *experience* of it. Many compatibilists believe that there is an unsuspicious notion of "could have done otherwise" that is devoid of any dubious libertarian underpinnings (Pauen 2004, 131). This notion is often brought in conjecture with George Edward Moore's analysis of being able to act otherwise as a counterfactual condition (Moore 1978, 150). According to Moore's analysis "being able to act otherwise" means that a person could have done otherwise *if* she had decided or chosen to do so. Moore's analysis is suggested to be compatible with determinism because it does not utilize any notions of contra-causal powers that add further difficulties in explaining human action. It is suggested to clarify the notion of "acting otherwise" in a metaphysically unsuspicious way that satisfies all demands for a reasonable theory of responsible agency. Persons are responsible for their actions because these had not occurred if they had not decided to perform them beforehand. Intelligible as this approach seems to be, it can be contested in at least three different ways. I believe that together these are strong reasons to reject Moore's analysis.

The first objection emphasizes that Moore's counterfactual analysis also applies (provides a true description) in situations where people were completely incapable to *actually* decide otherwise. Imagine the following scenario: If a mentally imparied person comes to the court and is told that he could have acted otherwise if he *had* decided so, then this might certainly be a true expression (Seebaß 1993a, 234). Analogously, it is clearly a true proposition that Iceland would not exist *if* the Eurasian and the American plate would not drift apart. The catch, however, is that Iceland had to emerge, because it is, regarding all that we know—hardly imaginable that this condition had not occurred—those plates had not drifted apart. Similarly, telling an insane person that he could have acted otherwise, had he decided to do so, just masquerades that he could *in fact* not have acted otherwise because he could not even have *decided* to act otherwise (Seebaß 1994, 226; Hartmann 2005, 5). Therefore, holding him responsible would be morally reprehensible. Moore's analysis only emphasises a necessary condition of free agency (the freedom to decide or freedom of will, so speak) while leaving unanswered the question whether this condition is ever fulfilled. A defender of Moore's analysis could immediately respond to this objection by acknowledging that this confusion is a result of counterfactual logic. A refined version of his analysis might be stated as follows: P acts in a certain manner, if P decides to act in a certain manner. This analysis is devoid of a counterfactual condition, but raises another libertarian objection. The condition to be fulfilled is that P decides to act in a certain manner instead of acting in a possibly different manner and a libertarian will be interested to hear whether P can *decide* to act in different manners: Whether he has the ability to act otherwise. Deciding to act differently is a necessary, but not a sufficient condition for free agency (Lehrer 1964). Thus, this second objection emphasizes that Moore merely *repositions* the questions of freedom from the moment of action to the moment of decision-making, and this move raises exactly the same concerns about agency in respect to a possibly determined world in which nothing else than what factually happens could eve happen (Seebaß 1993a, 233). Richard Taylor brings this objection to the point in a completely agreeable manner:

> It is said that, even assuming determinism, the necessary condition for responsibility is often fulfilled, for to say that an agent could have done otherwise means only that he would have done otherwise had he chosen to. But this neglects the fact that, if determinism is true, he could not have chosen otherwise. Indeed, by this kind of argument, one could say that, though a man has died of decapitation, he did not have to die, that he could have lived on – meaning only that he would have lived had he somehow kept his head on! And this is hardly the sort of contingency we want. (Taylor 2013, 309)

Finally yet importantly, a third objection deserves mentioning putting the finger on determinisms' implications for actually conceiving counterfactual conditions. After all, Moore's compatibilist analyses is a response to the challenge of determinism, and it suggests that, even if determinism were true, one could reasonably say that someone could have decided and then acted otherwise. To recall what we said about the characteristics of a deterministic theory: It suggests that the whole universe is governed by the necessary conjuncture of causes and effects. The universe is governed by the laws of nature, and this is an ontological thesis (Seebaß

1993b, 3). In fact, if determinism were true, there are only two possible ways according to which things could be different from how they are: Either the initial state of the world would have been different from the very beginning or the laws of nature were different from what they are. With the exception of religious presumptions, there is no way that any of these conditions could ever be true. Therefore, it seems that taking determinism seriously means acknowledging that nothing else could have ever happened differently, and no one could *decide* to do anything else than what one decides to do (Seebaß 2014, 224). Hence, this line of reasoning reveals the true difficulty of Moore's counterfactual analysis: The ontological nature of the thesis of determinism implies that counterfactuals of the sort that were employed by Moore can only have the status of thought experiments without any implications on our beliefs in the freedom of agents (Wolf 1990, 100). We can imagine many determined worlds that differ from ours by virtue of adhering to different laws of nature or by virtue of having different initial circumstances. In these possible worlds, people might "act" differently than in this one. However, in each of these worlds people cannot decide to do anything else than what they do because each of them is under the sway of the laws of nature—each in a different fashion—but all equally, heading towards an inevitable future. Conceiving possible worlds would not actually provide human agents in *this world* with more freedom if this world were determined (Seebaß 1993a, 242).

4.7   Utilitarianism and Retributivism

Before I return to compatibilism and genuine free will, we should detour a bit and explore a common misunderstanding concerning the above outlined theory of effect compatibilism. This while also advance our concept of responsibility as introduced in chapter three. It is suggested by a number of notable authors that because effect compatibilist notion of responsibility is allegedly primarily interested in influencing human behavior to the better, it will *always* find punishment appropriate when leading to a considerable effect regarding either the punished person's future behavior or regarding the reactive behavior of other members of society. This argument raises retributivist intuitions, and in its strong version, it charges effect compatibilism of being parasitic to notions of genuine free will. I will show that both objections—at least if directed to effect compatibilists like Schlick—are misled. Saul Smilansky enqueues Schlick confidently amongst promoters of "utilitarian theories of responsibility," and puts him in line with authors such as Hobbes, Mill, Sidgwick, Stevenson, and the early Alfred Ayer (Smilansky 2000, 27). In addition, Paul Russell, who proves in general to be a careful scholar of Hume's writings, argues confidently that Schlick "is willing to dispense with all retributive elements from punishment." (Russell 1990, 541) Furthermore, Jay Wallace, who calls the theory of effect compatibilism the "economy of threats account," considers this as a "utilitarian approach to punishment in the law." (Wallace 1994, 54–55) Wallace sees the classic application of the theory in Schlick's outline, and he suggests that Schlick's stance opposes retributivist ideas of punishment.

These suggestions are supported by passages where Schlick writes that "[punishment] is a natural *retaliation* for past wrong, ought no longer to be defended in cultivated society," and that "[…] punishment is an educative measure, […] to prevent the wrongdoer from repeating the act […] and in part to prevent others from committing a similar act […]." (Schlick 1962c, 152) He also says, "this problem is not identical with that regarding the original instigator of the act." (ibid., p. 152) And moreover, that it would be pointless to try to affect an insane person "by means of promises and threat," and this was a good reason for not charging such a person with responsibility (Schlick 1962c, 153). These suggestions seem to underpin the interpretation of Schlick as an anti-retributivist theorist. If it does not matter who the original instigator of an act is in order to motivate other people to omit similar acts, we feel pushed away from retributivist ideas of punishment. Regarding this approach, it appears to be acceptable to blame people for things, they have not brought about.

However, this extreme interpretation and the accompanied charges are part of a bigger misunderstanding. To resolve this, it is important to first note that Schlick had no interest in the study of normative ethics in the pursuit of answering questions as, for instance: "How ought we to live, or whom should we become?" His understanding of ethics opposes the view that sees ethics as a practice of *justifying* duties, obligations, dispositions, or other values as introduced in chapter one. Ethics in Schlick's understanding is a purely *descriptive* science that studies and explains human conduct: "The main task of ethics […] is to explain moral behavior. To explain means to refer back to laws: every science, including psychology, is possible only in so far as there are such laws to which the events can be referred." (Schlick 1962c, 144) In another passage, Schlick writes through ethical investigations "[…] we want only a simple determination of what, in human society, is held good." (Schlick 1962c, 89)

We should, therefore, understand his remarks about punishment, not as a justification of a utilitarian system of punishment—as, for instance, expressed in the formula: whenever the act of punishing someone maximizes the good, it is obliged or worthwhile to punish that person—,but as remarks about the *actual* system of responsibility ascription. Thus, there can be little doubt that in many contemporary states both deterrence and reformation indeed play a role in determining the degree of punishment (Rachels 1993, 132). More generally, morality—Schlick explains in the foregoing chapter—"is determined by the opinion of society, which is the lawgiver formulating moral demands. […] The content of the concept 'good' is determined in such a way by society that all and only those modes of behavior are subsumed under it which society believes are advantageous to its welfare and preservation […]." (Schlick 1962c, 96) This idea is purely descriptive and contrasts sharply with the idea of morality as an institution that justifies practical demands. In Schlick's descriptive ethics, acting morally right is equated with obeying the rules and laws a society believes to be most advantageous to its welfare, which depend on the societies' size, living conditions, surroundings, and so on (ibid., pp. 90–93). In addition, because of those divergences, we can explain the ethnographical and historical differences between the legal and moral systems of different societies: what is most advantageous for a society differs according to differences in their attributes. The *prescriptive force* of morality does not stem from its reasonableness but from societies pursuit to enforce it: "The state does in fact compel its citizens by imposing certain sanctions (punishments) which serve to bring their desires into harmony with the prescribed laws." (Schlick 1962c, 147) The "demand character" of morality stems from being told by others how "it is desired that we should act." (Schlick 1962b, 80) Both the moral code and the laws of a society are assumed to be determined by what legislators, parliaments, and the public consider as most conducive to the general happiness (ibid., p. 94). Only through this understanding of morality and law as the sets of duties and obligations that are most *useful* to society—considerations of a roughly utilitarian kind appear in Schlick's thinking, and these considerations are of a *purely descriptive* nature. Schlick believes that it is a matter of fact that "moral considerations" do not enter debates about legislation, but rather only considerations about what is most useful for society. Besides his general descriptive ethical stance, Schlick was a pronounced critic of utilitarianism as a normative ethics. He assailed utilitarianism as being "utterly inapplicable" because neither are the consequences of actions calculable, nor those consequences comparable (Schlick 1962c, 88). Schlick's opposition to utilitarianism is undeniable when, on the same page, he writes that it is a fundamental mistake to believe that "one can speak of the pleasure of different persons as of something comparable in magnitude."

Regarding this display of a classic anti-utilitarian reasoning, it becomes clear why it is misleading to confront Schlick with problems that are typically addressed to utilitarians. Given, that Saul Smilansky, for instance, asserts in such a manner that it would follow from Schlick's proposal (which Smilansky calls "effect compatibilism") in which people who are actually innocent will be (or must be) held responsible in a sanctioning manner as long as it is conducive to the general good. Smilansky argues that it might be beneficial regarding the general good to relax the "procedures for the apprehension of suspects, their prosecution and trial." (Smilansky 2000, 28) Doing so would probably have a positive effect on crime rates because it will result in fewer criminals escaping conviction and, thus, fewer criminals who continue to threaten others. It would make people feel safer, but it also brings about the possibility that more people that are innocent will be convicted. Overall, however, this might be more beneficial for society than alternative executive practices and, therefore, the best thing to do from a utilitarian point of view. Smilansky writes that "[only] in terms of justice, where there is something inherently wrong in the 'punishment' of any innocent person, is such a transformation a bad thing. And so only a non-utilitarian (or not-only-utilitarian) position could guard us from such injustice." (Smilansky 2000, 29)[55] It is, of course, a fore-going empirical question whether one can assess such a general transformation of the mechanisms concerning the executive branch of a country in utilitarian terms and whether the outcome of such assessment would indeed equal the results Smilansky suggests. The notion of "being beneficial" is fairly wide and one can express doubts about the long-term effects of such measures—the erosion of trust might be a long term effect of such transformation, and this would be an extremely negative result that has to be accounted for in the assessment of such measures. Thus, the utilitarian could argue that factually is *not* beneficial to punish the innocent in this case by making the executive processes more stringent because it would excavate people's trust in the legal justice and the legitimacy of legislation. Moreover, there might be other forms of utilitarianism that are not threatened by such an example.

These "technical" issues should not be discussed any further at this point. It is more important for the present discussion that we can propose a straight-forward response in the spirit of the above outlined descriptive moral theory to Smilansky's challenge: Schlick can argue that he does not even aim to *justify* the occasional discrepancy between holding someone responsible (which on Schlick's basis is a person who felt free and upon whom the motive has acted) and sanctioning her with blame or punishment as a means of deterring others or reforming her. For Schlick, the existence of punishment for such means is a descriptive fact about our (European and Northern American) societies. It is a factual aspect of our "cultivated" social reality that sometimes punishment is decoupled from having felt free or being the right point of application of the motive. This occasional discrepancy is merely a *descriptive fact*, and there is evidence that we indeed sometimes de-

---

55   The literature on retributive justice is vast, and it should be mentioned here that there are authors who argue that retributivism is far from being the fair idea that Smilansky suggests. Note that retributivism makes punishment look like cruel and blind vengeance, and as something intrinsically good because it does not fulfill any purpose (Wallace 1994, 61). Such a stance appears to be similarly wrong. My intuition is that both extremes—the retributivist and the anti-retributivist—are one-sided. Both do not do justice to the variety of our normative convictions. This is in accordance with my virtue ethical standpoint.

couple the sanctioning of people from their actual responsibility. Politicians some-times argue publicly that—often following a widespread social demand—one has to take hard measures to unmistakably display that such and such behavior cannot be accepted. This has, for instance, been the case when former American President George W. Bush justified the first wave of the so-called "War on Terror" which led the United States to invade Iraq. "Just desert" has certainly not been the sole mo-tive behind this operation, since the terrorists that attacked the World Trade Centre were killed during the attack (except for leading figures like Osama bin Laden). The operation *also* aimed at proving that one should not play games with a military superpower and, thus, to prevent future terrorist attacks *through intimidation* (a project that failed drastically). Moreover, there are more ordinary cases in which the practice of blaming and punishing diverges from our convictions about the responsibility of a person. Imagine, for instance, a teacher who gives a lecture and during her teaching a number of students (adults) in the front and the back rows talk to each other and disturb her lesson. After telling them repeatedly to be quite, she randomly chooses a girl from the front row and asks her to leave the classroom. She knows that she cannot suspend everyone, but that silence will be reinstalled, if she suspends at least one of them to display who the boss is. Clearly, she is aware that the girl picked can claim that this is unjust: She was not the only one talking, and her behavior alone would not have disturbed the course—only together did the students produce a disturbance. But for the teacher, the end seems to justify the means and one can also say that this was a reasonable thing to do as an educa-tive measure.[56]

---

56   Angela Smith discusses a number of great examples such as this one: "From the
      fact that I do not 'hold' someone responsible for an objectionable attitude or action,
      in the sense of actively blaming her for it, it might be inferred that I do not hold her
      to be responsible for that thing, in the sense of open to moral appraisal, or that I do
      not hold her to be culpable for it, in the sense of open to legitimate moral criticism.
      But these things simply do not follow. If a good friend of mine is under a lot of
      stress, for example, I may not 'hold' her responsible, in the sense of actively blam-
      ing her, for some insensitive comments she makes to me. I can judge both that she
      is responsible for her comments, and that she is open to legitimate moral criticism

Therefore, it can be said that *sometimes* the proportion of our sanctioning reactions is decoupled from actual responsibility, and that it can *sometimes* (to silence the class, for instance) be reasonable to do so (Wallace 1994, 60). In other words: A person can be morally responsible without being blamed or punished and *vice versa*—a person can be blamed without being (at all or at least to the same extend) blameworthy (Smith 2007). The *adequacy* of sanction and blame is a question of normative ethics. Since I neither aim to provide a full-fledged moral theory, nor do I believe that conditions determining unambiguously when disproportionate punishment can be allowed or (even required) enumerated, I grant myself to remain unspecific mentioning only few examples. I also suggest that the factual occasional discrepancy between actual bearing of responsibility and sanctioning for the sake of deterrence and reformation is the root of the paradox of consequentialist moral luck, which will be discussed in the next chapter. Instances of consequentialist moral luck reveal that we sometimes hold people responsible in terms of blame and punishment for things that have been beyond their control, and this challenges our intuitions about actual responsibility. However, even if it is true, as asserted before, that we sometimes, as a means of deterrence, reformation, and retribution, detach the proportion of sanction from the degree of a person's wrongdoing, an ethical theory that entertains this separation *universally* cannot be acceptable and this seems to be the peril of utilitarianism. Thus, it seems clear to me—without being able to provide conclusive evidence within the scope of the present discussion—that utilitarians, more often than most of us, would find this appropriate and will suggest or require the application of sanctions to innocent people or allow for disproportionate punishment to affect the general happiness, welfare, or whatever value the utilitarian finds worthy of being maximized. The

---

for them (because they are hurtful). But given the circumstances, I may decide that it would be uncharitable for me to take up attitudes of anger and resentment, or to explicitly reproach her in any way. In making such a judgment, however, and in renouncing these attitudes and responses, I need not thing that my friend is not really responsible, or not really at fault, for her behavior." (Smith 2007, 470) Overall, her reasoning has many similarities with my own approach. However, I believe that the considerations that affect our practice of holding people responsible exceeds the three mentioned by her, which are the blamers' relation to the doer, significant of fault, and the agents' dealing with their wrongdoing (ibid., pp. 479–482).

literature is full of striking examples and thought experiments (Rachels 1993, 115; Wiggins 2006, 213). It is an essential aspect of utilitarianism that there are no absolute values such as justice (in terms of appropriate and proportional desert) or dignity. Therefore, any value can in principle be undercut, if the stakes are high enough, whatever "high enough" means in particular circumstances (Wiggins 2006, 214).

It is important to note that the utilitarian can argue against this by entertaining a distinction between *believing* someone is responsible and *holding* someone responsible (Watson 2004d, 267). Believing that someone is responsible might then be appropriate when the following biconditional is fulfilled: P is responsible if and only if, P has balanced reasons and pursuit ends that have been formed by such foregoing reasoning. The utilitarian might add: This distinction between *believing* in someone's responsibility and *holding* her responsible through articulation or other practical means can be upheld even though in reality, they often divert—they divert as often as the contexts of human action contingently demand. I think it is sound to employ this strategy of accounting for the *occasional* discrepancy between practices and articulations of responsibility as desert and punishment, and the beliefs in responsibility as prevalent in common sense morality. I am inclined, however, to assume that utilitarians have to accept that believing and holding responsible can *in principle* diverge depending on the external conditions that influence the external *consequences* of those sanctioning reactions and to argue that such divergence in principle comes with a high cost (Nagel 1991a). It is hard to see, how one can permanently uphold the distinction between believing in someone's responsibility and, at the same time, accepting that the treatment of this person is *always* merely a function for promoting the general welfare. My suggestion is that entertaining such divergence in principle presumes (or eventually leads to) severe alienation from social commitment, hypocrisy, and psychological dislocation (Chappell 2009, 73).[57] These suggestions are not telling. They underline my

---

57   Chappell charges utilitarianism with this verdict for employing an analogous strategy concerning the public-ness of morality. Utilitarians sometimes claim that they could permanently separate believing what is the right thing to do from what is

doubts regarding the assertion that punishment and other sorts of sanctions should *in principle* be detached from actual wrongdoing, although I have explicitly suggested that such detachment is *sometimes* reasonable. From the perspective of common sense morality, we suggest in contrast to the utilitarian that in some exceptional cases such detachment can be *permissible*. The previous reasoning has dealt with questions of normative ethics and it must be clearly emphasized again that none of these concerns must bother a descriptive ethicist like Moritz Schlick.[58] Regarding Schlick's overall agenda, the normative question of retributivism concerning the proportion and appropriateness of punishment makes no sense. Neither is the retributivist challenge a threat to modern, elaborate forms of effect compatibilism like Vargas' revisionism (Vargas 2007, 157). Revisionists do not have to commit *ex ante* to a normative ethical standpoint, such as utilitarianism, or, more generally, consequentialism. Revisionism merely holds that we assess people by the "norms we accept"—and these can be pluralist, consequentialist, deontological, or whichever normative ethical theory turns out to be most widely accepted (ibid., pp. 158–159).

---

publicly articulated to be the right thing to do. This must be sometimes separated because articulating what is the right thing to do can have suboptimal consequences—since the right thing to do can be something so demanding that it will discourage people from even attempting to comply. Singer discusses this possibility as a problem of moral motivation in a chapter of his Practical Ethics (Singer 2008, 242–246). Thus, Timothy Chappell asks, irritated concerning this proposal: "Who are these people who understand that our common moral life is based on falsehoods, but also participate in that common moral life? How do they manage to do this without hypocrisy, irrationality, alienation, or other severe psychological dislocation?" (Chappell 2009, 73) I could not express it any better.

58 Schlick's theory has clearly other weaknesses, weaknesses that are usually associated with relativist positions: If one sees morality solely as the (arbitrarily) established rules and norms of a society, it is hard to explain the nature of moral disagreements within these societies and across different societies. Can one make sense of such deeply rooted intuitions about the wrongness of discriminating minorities whether they live in Iraq, China, or Germany? Relativism cannot make sense of disputes concerning discrimination across these societies because it denies the existence of a shared object of conflict—namely that there is something that is the right thing to do or the right attitude regarding discrimination that spans across social boundaries (Williams 1997, 16–17; Singer 2008, 6).

Let us return to the topic of compatibilism and freedom considering this issue has not been settled in the previous exploration. The argument against punishing the innocent is sometimes put forward as a libertarian or incompatibilist challenge to the compatibilist approach of holding responsible. Therefore, it is most important for the present purpose to ask what exactly does Smilansky mean when talking about a person as "innocent." Does a debate about such concept of "innocence" even *concern* the issue of compatibilism and libertarianism that has bothered us in the previous sections? I have been using the term "actual responsibility" several times in the previous reasoning without being very specific about this concept and I also argued that in common sense morality we sometimes distinguish between (our belief in) someone's responsibility and someone's being reasonably held responsible for educational, deterring, or reformatory ends. This is not only supported by cases in which we divert from actual responsibility for the sake of deterrence (as in the classroom case), but also complementarily when we refrain from articulating resentment or reward despite facing "actual responsibility." We sometimes do believe a person is responsible, but do not react at all because the action has been too trivial (Watson 2004d, 265; Braham and van Hees 2013). For example, when someone gets up from a chair intentionally (to have a break or take a deep breath), we believe that she is responsible, but we do *not hold* her responsible in a rewarding or sanctioning manner. As Angela Smith writes with regard to another example: "[In] some cases the failing may be so trivial as to warrant little or no criticism at all." (Smith 2007, 480)

It seems that Susan Wolf develops an argument in retributive spirit that suggests an open flank in compatibilist thinking and seems to support a libertarian standpoint. This argument might be applicable even if compatibilists approve of the distinction between believing in someone's responsibility and holding responsible, as suggested above. The compatibilists concept is, thereby, accused to ride piggyback on a notion of genuine free will. Wolf writes:

> It is essential to the nature of resentment that it can be deserved or undeserved, appropriate or inappropriate, where the conditions of desert or appropriateness go beyond the establishment of the fact that the person **really** did perform the action in question and that the action **really** did, or would have been expected to do, some harm to us [own emphasis]. (Wolf 1990, 20)

Thus, Susan Wolf's argument against effect compatibilism can simply be put in the form of a *modus tollens* that goes: 1) If effect compatibilism is true, then it is appropriate *to sanction* X, if X is the bearer of the motives or desires that led to φ (or if X has in a weak sense intended φ, or if X felt free to do φ)[59], 2) It is *not* appropriate to sanction X, if X is the bearer of the motives or desires that led to φ *and* X is *not* genuinely free, therefore, 3) effect compatibilism is wrong.

As argued before, in general I think that there are good reasons to apply the suggested distinction between believing in the responsibility of a person and holding them responsible, and it seems to me that Schlick can perfectly employ this strategy. Thus, Schlick might argue that someone is responsible—or, in other words: We can believe in her responsibility—when this person has been devoid of external compulsion, when they "felt that she could have acted otherwise," and when they have knowledge about having "acted on their own desires." Regarding this modification, which is, as I believe, completely in harmony with Schlick's thinking, Susan Wolf's argument can be modified to say in short: 1) If effect compatibilism is true, then it is appropriate *to believe* in X's responsibility, if X is the bearer of the motives or desires that led to φ (or felt uncoerced and free in φ-ing), 2) It is *not* appropriate to believe in X's responsibility, if X is the bearer of the motives or desires that led to φ *and* X is *not* genuinely free, therefore, 3) effect compatibilism is wrong. Thus, what Susan Wolf presumes in the second premise in this slightly modified (and strengthened) interpretation of the argument is the requirement of genuine freedom to even appropriately *believe* in someone's responsibility. Remember that sanctioning and desert are not always appropriate or deserved in common sense morality and, thus, believing and holding responsible should not be equalized even if they do usually coincide. In short: It is obvious that the requirement to settle this question is a compelling argument for when we are rightly allowed to *believe* in someone's responsibility and it is here that the question about genuine free will returns: Is genuine free will a necessary condition to believe in someone's responsibility (Campbell 1951, 27)? Schlick and other

---

59   For the current purpose it is of secondary interest for this reconstruction whether Schlick thinks those conditions have to be conjoined, even mean the same, or are rather disjunctively necessary to be morally responsible.

compatibilists deny this and argue that something less strong suffices for being responsible, which is being free of external compulsion and having the feeling of being uncoerced. This is expressed in the second conditional in first premise. In contrast to Schlick, Susan Wolf claims in the above cited passage that judging someone responsible makes sense only if the person *really* did perform the action in question, which is the case if and only if they could have freely chosen the motives according to which they acted. However, this move cannot convince: Plainly asserting the truth of the second premise begs the question of the free will debate. The arguments confronted here *presuppose* that either strong free will or something much weaker are necessary to be appropriately considered as being responsible and, therefore, must both be charged of a *petitio principii*. One cannot summarize this objection any better than Jay Wallace did:

> It would prejudge the debate in a different way, however, if one built into the stance of holding people responsible the belief that the targets of that stance have strong freedom of will. To do so would over intellectualize our practices, making incompatibilism virtually an analytic consequence of what we are doing when we hold people morally responsible, whereas we actually tend to adopt this stance without entertaining any clear ideas about whether those at whom the stance is directed have strong freedom of will. (Wallace 1994, 60)

Because of the validity of this objection against Wolf's argument, I believe that it is much more promising to defend libertarianism in the manner presented at the outset of this chapter. Assuming that someone is responsible and holding that person responsible presumes a notion of *agency*, which already implies that a person performs actions willingly while she could have done other things instead. The basic idea of agency is already incompatible with determinism. Schlick and other compatibilists cannot meaningfully speak of agency in this way *without* presupposing libertarian assumptions. This conceptual objection to compatibilism does not rely on a normative notion such as appropriateness or fairness.

4.8   Conclusions

A longer detour was necessary to do justice to the complexity of the vast issues that emerge in the triangle freedom, determinism, and responsibility. In the previous chapter, I discussed visioneering as a case about a type of social agent that supposedly influences innovation processes by providing visions and narratives about the future. With regard to this example, it was argued that actions are to be distinguished from mere behavior in virtue of being intentionally performed, and since epistemic limitations enclose the scope of things, one can intend or willingly accept as possible effects of one's actions. What did not appear to be problematic at all was that within this scope of epistemic limitation there are actions, which clearly fulfill these criteria (if only of a self-forming type) and for which visioneers can be clearly held responsible. However, in pursuit of explaining actions like "influencing innovation dynamics" or "training virtuous behavior" and taking into account humans belonging to the natural world, it becomes less and less comprehensible how people can intend and do things while other natural processes seem to be under the sway of natural laws that necessitate their behavior. It was important to emphasize at the outset that our lack of doubt about such notions in ordinary contexts cannot satisfy a deeper philosophical interest in understanding agency and the menace of determinism. However, as pointed out, determinism is in itself an idea that lacks a strong fundament. Given that the most robust knowledge from the natural sciences is gained by shielding observable phenomena and trying to reproduce their effects in laboratories, it is even regarding the most obvious candidates for natural laws not evident in which sense one can say that they describe *natural* regularities. Determinisms implication that such laws govern the course of the whole universe—which is an even stronger claim—can hardly be considered as a plausible theory. Agency, however, is also threatened by causal theories that emphasize the event ontological substructure of actions and the causal relations between those events. However, by pressing actions into a causal corsets the nature of agency is deprived of its central feature: the doer's *active* pursuit of certain ends. This is made intelligible by the reasons agents have that represent the results of deliberations and that have more or less thoroughly been undertaken before a decision has been made. The existence of this phenomenon is in fact a good reason to doubt determinism's plausibility, which aspires to provide a comprehensive description of what is *happening* in the universe.

Compatibilists doubt that genuine (undetermined) agency is an intelligible idea, and, at the same time, they assert for a variety of reasons that there are still good grounds to continue speaking of people's responsibility. They either refer to the ordinary experience of decision-making during which we normally do not perceive any kind of resistance or opposition (except for the authority of and persuasiveness of good reasons (Kant 1996, 41)). Furthermore, some effect compatibilists argue that it is reasonable to continue reacting with praise, blame, and other incentives or sanctions to people who acted intentionally or—Schlick's words—to "the 'doer' is the one upon whom the motive must have acted in order, with certainty, to have prevented the act." (Schlick 1962c, 152–153) Both arguments have been confronted with substantial objections. The familiar phenomenon of "acting intentionally" might as well be an illusion in a determined world. Entertaining the idea that intentions can be attributed to people as the ends, these persons have set-up for themselves implies the possibility that they could have chosen different ends. This is not denied by compatibilists; however, its significance is underestimated or relocated from actions to decisions as in Moore's conditional analysis. This move cannot solve the difficulty that there is only one future in a determined world and talking about alternative possibilities is, therefore, purely *imaginative*.

Regarding the issue of retributive justice, we have seen how closely the descriptive and the normative aspects of responsibility are intertwined. It is clear that the assumption only genuinely free human beings can be appropriately held responsible and addressed with sanctions or blame begs the question of genuine freedom in favor of a libertarian viewpoint. Jay Wallace puts this correct objection forward against Susan Wolf's argument. As argued before, effect compatibilists like Schlick do not have to determine the conditions under which it would be appropriate to influence someone's behavior to the better with praise, incentives or on pain of penalties. As a response to cases where holding someone responsible does not coincide with that person meeting the standard criterions of responsibility, as pointed out in this and the previous chapter (alternative possibilities, intentionality, authorship condition), effect compatibilists can entertain a distinction between being responsible and being held responsible. This reinforces the burden of proof on the effect compatibilist to justify why one can reasonably *believe* in anyone's responsibility in a determined world. Furthermore, it reminds us of the unsettled task for normative ethics to provide the conditions according to which sanctions, blame and punishment can be considered appropriate and be applied even if someone is not responsible. Such normative issues also arise when we focus on cases in which people are punished for things that have been beyond their control; here too our beliefs and our reactions come into conflict. In the problem of moral luck, which will concern us in the next chapter, issues of responsibility and determinism continue to coincide.

5. Moral Luck and Intelligibility

5.1 Moral Luck

The last chapter explored some issues regarding the metaphysical underpinnings of attributing responsibility and the concept of agency. However, there are still a number of substantial questions to be answered in this area. Chapters two and three dealt with the framework conditions in which innovations are being made and the epistemic limitations due to complexity, which social actors like visioneers and innovators that aspire to develop new technologies, faced in these contexts (Seebaß 1993a, 226). There are visible *external constrains* and *contingencies* in producing novelty, and my preliminary conclusion in these earlier chapters showed these external constrains do not undermine agency. But how strong are these constrains really, and whether and how do we take them into account in moral evaluations? Do we consistently take the contingencies into account that prevail in such complex environments affecting the consequences of actions, and the unequal distribution of obstacles that people face in their (professional) lives when evaluating their actions? In regards to the problem of moral luck, the issues about evaluating and adequately reacting to actions that might be the *results* of contingent events, which might *cause* similarly contingent events coincide.

Therefore, we should discuss the challenging viewpoints of Thomas Nagel, which have been developed in an article called *Moral Luck* (ML) and similarly in his book *The View from Nowhere* (VFM) (Nagel 1986, 1991b). Nagel's arguments have received considerable attention by a number of notable thinkers in philosophy (Andre 1983; Strawson 1985; Mele 1999; Waller 2011). However, the problem of moral luck is also frequently discussed in the context of RRI which again underlines its significance for this the present enquiry (Danneels 2004; Grinbaum and Groves 2013; Stilgoe et al. 2013). At the end of this chapter, I will transfer the insights of my discussion to the innovation context. The problem of responsibility is understood by Nagel as the problem of the freedom of others while the problem of autonomy is understood as the problem of one's own freedom. Nagel discusses a number of arguments that together highlight the fact that by trying to understand

© Springer Fachmedien Wiesbaden GmbH, part of Springer Nature 2018
M. Sand, *Futures, Visions, and Responsibility*, Technikzukünfte, Wissenschaft
und Gesellschaft / Futures of Technology, Science and Society,
https://doi.org/10.1007/978-3-658-22684-8_5

the idea of responsibility as much as the notion of autonomy, we are sooner or later forced to view ourselves from an external perspective. This "external perspective" is conceived in Nagel's texts as an event ontological point of view on the world. Such a point of view on the world excludes agency, which is, the standpoint with which we are normally most familiar. From the internal standpoint, the future looks open to us. We can choose to instantiate possible futures by choosing to do something. However, from the external standpoint, the world appears to consist merely of series of events. From this external perspective it seems that there is no place for agency in the world: "The same external view that poses a threat to my own autonomy also threatens my sense of the autonomy of others. This in turn makes them appear as inappropriate objects of admiration and contempt, resentment and gratitude, blame and praise." (VFM, p. 112) Thus, the internal and external standpoints are essentially incompatible, which creates a tension that cannot be overcome. Nagel writes in *Moral Luck*:

> A person can be morally responsible only for what he does; but what he does results from a great deal that he does not do; therefore he is not morally responsible for what he is and is not responsible for. (This is not a contradiction, but it is a paradox.) (Nagel 1991b, 34)

Later in the text, Nagel tries to understand the nature of this paradox, which he describes, on a more abstract level. In *Moral Luck*, when Nagel states that people are factually held responsible for many things that "result from a great deal that he does not do," he refers very generally to aspects of decision making that are, according to him, beyond the agent's control. He names the following different types: circumstantial, constitutional, and consequential moral luck. More generally, he argues that from an external perspective it looks as if the agent—his actions and character traits—are "swallowed" up by the order of mere events. Nagel elaborates further on this rather abstract perspective on the agent and his actions in VFN. Before we consider his argument in the book and start discussing his claims, we should have a closer look how he finishes his article:

> The problem arises, I believe, because the self which acts and is the object of moral judgment is threatened with dissolution **by the absorption of its acts and impulses into the class of events**. Moral judgment of a person is judgment not of what happens to him, but of him. It does not say merely that a certain event or

state of affairs is fortunate or unfortunate or even terrible. It is not an evaluation
of the state of the world, or of an individual as part of the world. We are not
thinking just that it would be better if he were different, or did not exist, or had
not done some of the things he has done. We are judging **him**, rather than his
existence or characteristics. The effect of concentrating on the influence of what
is not under his control is to make this responsible self seem to disappear, swallowed up by the order of **mere events** [own emphasis]. (Nagel 1991b, 36)

This effect—as Nagel argues in VFN—of the disappearance of the active self can
be brought about by watching ourselves from the outside. In his book from 1986,
his argument focuses solely on the abstract opposition between the external or objective standpoint, which "opposes" the internal standpoint. Here, Nagel no longer
refers to the examples of constitutional, consequential, and circumstantial luck,
which he discusses in his article originally published in the *Proceedings of the
Aristotelian Society* in 1976 and reprinted in his *Mortal Questions*. It is not clear
whether he was no longer convinced that the examples brought forward in *Moral
Luck* could substantiate or support the claim that the two perspectives together
give rise to an unresolvable paradox, or whether his more generally introduced
idea of the opposition of the internal and the external perspective can substantiate
his viewpoint all by itself. Before we discuss the arguments in detail let me cite
another passage from VFN that resonates with the central claims of the above-
cited passage from ML, and shows the connectivity between the two pieces:

> Something peculiar happens when we view action from an objective or external
> standpoint. Some of its most important features seem to vanish under the objec-
> tive gaze. Actions seem no longer assignable to individual agents as sources, but
> become instead components of the flux of events in the world of which the agent
> is a part. The easiest way to produce this effect is to think of the possibility that
> all actions are causally determined but this is not the only way. The essential
> source of the problem is a view of persons and their actions as part of the order
> of nature, causally determined or not. That conception, if pressed, leads to the
> feeling that we are not agents at all, that we are helpless and not responsible for
> what we do. Against this judgment the inner view of the agent rebels. (Nagel
> 1986, p. 110)

This passage is a clear indicator that Nagel did not fundamentally revise his position despite disposing of the particular examples of moral luck that he discusses
in ML. First, it is important to mention that Nagel does not defend the thesis that

autonomy and responsibility are mere illusions. Autonomy is an intelligible idea, according to Nagel, that has its foundation in the internal perspective of the agent to whom the future is open: "[…] when we act, alternative possibilities seem to lie open before us: to turn right or left, to order this dish or that, to vote for one candidate or the other—and one of the possibilities is made actual by what we do." (Nagel 1986, 113) However, it is the thesis that this internal and external point of view are naturally sooner or later adopted *and* are incompatible that produces this inevitable tension.

First, it is important to understand the reasons for assuming that there is an inescapable pull to the external standpoint. The reasons can be roughly diverted into two larger classes. The first class contains the instances of consequential, circumstantial, and constitutive moral luck that contributes to the feeling that there are a number of things beyond the agent's control that naturally give rise to viewing him as a part of a series of events. These are outlined in greater detail in ML and should be discussed separately, diverted again in two subgroups: the first contains consequentialist moral luck.[60] These instances point to aspects that concern the circumstances or the surroundings in which agents act and the respective consequences affected by them that are beyond their control and supposedly affect our evaluation of the agent's behaviour. The second class of instances are cases of constitutive moral luck which concern the agent's character traits that are also not consciously chosen but rather given by birth. In contrast, the second class of "arguments" that give rise to the paradox is closely connected to Nagel's concept of the external and internal perspective. In VFN he describes in more detail what he understands as a "perspective." He explicitly argues, in this case, that the problems of autonomy and responsibility are not "verbal problems" but rather "a bafflement about our feelings and attitudes" that are conjoined "with a loss of confidence" in our beliefs in freedom and responsibility (Nagel 1986, 112). It is, therefore, reasonable not to treat this "argument" like a normal argument, which can be tested on validity or logical consistency. Instead, I will seek to understand the thesis of

---

60   For the sake of length and simplicity, I will neglect circumstantial moral luck also mentioned by Nagel.

paradox and see how significant the supposed bafflement of feelings and insecurity about freedom and responsibility really are after thorough consideration. Closely related to the problem of perspective is another issue that appears in VFN, which is known as the intelligibility problem, or in Nagel's approach as the "explanatory gap." The explanatory gap also pushes the agent to the external perspective and, thus, creates again the bafflement of feelings and the tension between the internal and external perspective.

I think that each of these problems demand distinct responses, and I will show that one can indeed propose satisfying responses to each of them. The most complex problem that Nagel brings up—the intelligibility problem—will be discussed at the end of this chapter. Let us first start with the idea of consequential moral luck. Consequential moral luck is particularly interesting for the present consideration because this kind of reasoning also is advocated in the debate about RRI. Alexei Grinbaum and Christopher Groves mention the issue of consequential moral luck as particularly striking in the area of innovation characterized through division of labor and opaqueness (Grinbaum and Groves 2013). As mentioned before, innovators do not control the large parts of the system that influence the innovation journeys (Rip and Voß 2012). Hence, the outcomes of their innovative actions are largely a matter of good fortune. If the circumstances are right and an innovator joins an ambitious team with creative colleagues, attracts some well-meaning donors, and finds a market niche, he could become a leading figure in the annals of innovation. Otherwise, his motivation might seem quixotic and his efforts will lack appraisal. Nevertheless, many of these contextual aspects are not in his control, but they largely determine the success of his endeavor. Paradoxically, in our evaluations of actions we often seem to forget about the arbitrariness of the numerous circumstances that together contribute to the success of actions. This is also the general idea that Nagel outlines in his discussion of consequential moral luck. As mentioned before, the problem of consequential moral luck is an instance of a wider problem, which is that the agent is from an external point of view swallowed up by series of events. Consequential moral luck pushes us to view ourselves from the outside. In the following passage from ML Nagel connects the problem of consequential moral luck and the wider problem of autonomy:

> The inclusion of consequences in the conception of what we have done is an acknowledgment that we are parts of the world, but the paradoxical character of

> moral luck which emerges from this acknowledgment shows that we are unable to operate with such a view, for it leaves us with no one to be. [...] Once we see an aspect of what we or someone else does as something that happens, we lose our grip in the idea that it has been done and that we can judge the doer and not just the happening. This explains why the absence of determinism is no more hospitable to the concept of agency than is its presence - a point that has been noticed often. Either way the act is viewed externally, as part of the course of events. (Nagel 1991b, 38)

Let us consider the standard example to better understand the problem of consequential moral luck—luck in "the way things turn out":

> If someone has had too much to drink and his car swerves on to the sidewalk, he can count himself morally lucky if there are no pedestrians in its path. If there were, he would be to blame for their deaths, and would probably be prosecuted for manslaughter. But if he hurts no one, although his recklessness is exactly the same, he is guilty of a far less serious legal offence and will certainly reproach himself and be reproached by others much less severely. (Nagel 1991b, 29)

Nagel emphasizes this example for contrastation with our ordinary convictions. Our ordinary conviction about responsibility attributions is that we usually aim for an evaluation of a *person*—the agent themselves—when morally assessing actions (ibid., p. 25). In ordinary moral judgments, we assume that it is not plausible to hold people responsible for "what is not their fault, or for what is due to factors beyond their control." (p. 25) Here, Nagel outlines the case of the drunken driver as an instance of luck entering into our moral evaluations. In one of the outlined course of events, the drunk driver was lucky not to have harmed anyone. Nagel assumes that our ordinary standard is to treat him as a reckless and blameworthy person. However, a few apologetic words and the event will soon be forgotten. Now consider a very different possible course of events including some pedestrians who cross his path and the drunken driver kills all of them. The application of our "ordinary standards," as Nagel says, results in a much more severe reaction possibly ending one's friendship with him or abandoning him from social circles. The legal responses would also be much more devastating, most likely including a prison sentence for negligent homicide. Here, Nagel reminds us that in both cases the negligent act of having driven drunk is exactly the same, but our evaluations differ substantially, and it seems that in these evaluations we take into account things that are ordinarily excluded from our judgments—aspects that are beyond the agent's control—such as whether there are people on the street or not. We recognize that we are (and hold others) responsible for things we could not control, although we simultaneously believe that we are only responsible for the things that were within our control. This is the paradox that has been mentioned before. Nagel identifies the analogy between this case of luck regarding the way things turn out with Bernard Williams' more or less fictional example of Gauguin (Williams 1981a, 22–23). Williams, who also discusses an example involving a negligent driver running over a child, considers the problem of moral luck as an incoherence in our moral thinking in contrast to Nagel's phrasing of the problem as a "paradox." He outlines the case of Gauguin, who neglects social obligations towards his family and gives up his decent life as a banker to pursue becoming a painter. Williams conceives Gauguin not as a morally insensitive person, but as someone who is aware of the moral implications of his decision. He makes a conscious decision based on extreme uncertainty regarding its outcomes. Williams argues that the "the only thing that would justify his choice will be success itself." (ibid., p. 23)[61] Some of the aspects that determine the success of his decision to become a painter are extrinsic (like staying healthy and finding willing patrons), and some are intrinsic (his talent as a painter) in Williams' opinion, which coincides with Nagel's distinction of constitutive and consequential luck. Both are not entirely within the agent's control.

The title "consequential moral luck" seems to suggest that ethics, which evaluate actions mainly (or solely) in light of their consequences, suffers most from this problem which is also how Grinbaum and Groves interpret the problem of moral luck; as a threat mainly for consequentialism (Grinbaum and Groves 2013, 125–126). Williams indeed directs the case of Gauguin against utilitarianism as the most famous type of consequentialist ethical theory but also against rule-based ethical theories (Williams 1981b, 24). Given the uncertainty of the outcome, Williams argues, Gauguin could not have justified his decision from a utilitarian point

---

61   When hearing this example, I cannot resist thinking counterfactually about the possible adjustment of our moral evaluations had Edward Snowden's revelations led to a world war, which thankfully has not been the case.

of view at the time he had to make that decision. Since Utilitarians cannot deny the value of Gauguin's art, their theory could not have provided the justificatory resources to defend his decision when he made it. At the time of decision-making, the *direct* and certain neglect of his family stood against the extreme *uncertainty* of artistic success and fame as legacies. Success is an *ex post* state of action, occurring when decision-making and justification have already taken place (ibid., p. 25). In fact, such fundamental decisions regarding one's own biography are not usually made with balanced alternatives and their accumulated outcomes against each other. I will return to this aspect in chapter seven when I discuss the existential pleasures of innovators and argue similarly that the value of living a meaningful life cannot be captured by ethical theories that reduce the good to a single aspect, such as generalizable duties or outcomes (Wolf 2010). Virtue ethics—as I will argue then—is a plausible candidate for an ethical theory that can deal with the varieties of incommensurable values (partial and impartial) and other worthwhile aspects, such as one's meaning in life (Nagel 1991d). This layer of critical moral reasoning is not dominant in Nagel's text, but it is an essential aspect of Williams' approach to the theme of moral luck. In contrast to Williams, Nagel assumes that the problem of moral luck mainly afflicts the system of our ordinary moral beliefs and reveals a general paradox of our responsibility ascriptions. In general, it seems reasonable to assume that, besides ordinary morality and consequentialism, any ethical theory in which the consequences of actions play a role—even if only a derivative or minor manner, such as in virtue ethics and particularism—moral luck threatens a coherent picture of responsible agency. The arguments that will be provided in this section presume a general standpoint and can, thus, be adopted by advocates of different ethical theories. Thus, is there a way to resolve the paradox?

Judith Andre pointed out two suggestions in a critical response to Williams and Nagel (Andre 1983). First, given a certain distance to the case of the drunk driver and allowing for *proper reflection* on the agent's responsibility, the distance between those initial reactions that opposed each other seems to starkly diminish. Nagel presents the case by outlining initial reactions that also coincide with our spontaneous evaluation of the case. He presumed, thus, that there would be a crucial opposition in our attitudes regarding the driver that hit the pedestrians, and the driver that was lucky enough to make it home safely. However, after thoughtful consideration, we will indeed make him aware that he could have killed a number of innocent people. This is an expression of a real resentment and the disappointment about his misconduct even if in fact nothing happened.[62] For many of us, a simple apology is not a sufficient reaction for putting oneself and other people in life-threatening danger as the negligent drunk driver did. For some of us accepting such existential risk can even cause more severe reactions, such as ending a friendship. This means, that the offences in the case of the lucky drunk driver can be severe. On the other hand, when we realize that there is really no difference regarding wrongdoing in the case of the person that hit the pedestrians, we will pity him for his misfortune instead of blaming him for his misconduct. It is more likely that we are rather sorry for the unlucky driver after proper consideration. Held legally liable and probably punished with a prison sentence, this person will probably suffer from this and his bad conscience alike. Many of us believe that no further offences are required as a moral reaction. This shows that those initial diverging reactions can *converge after reflection* (Andre 1983, 203).

Second, there is an understanding of responsibility that is, in the words of Judith Andre, more "prosaic": "[T]o be responsible is to have an obligation to rectify bad consequences. If I break your vase I must replace it. I can be responsible in the second [rectification] sense without being in the least blameworthy, although often the two coincide." (Andre 1983, 205) I understand this point as a reminder of my reasoning in the previous chapter where I dwelled on the distinction between being responsible and holding someone responsible: our (moral and legal) *reactions* to misconduct accord with a variety of different purposes. To blame for wrongdoing, in a sense that is adequate to misconduct in order to encourage better behavior, is only one of the possible *purposes* of moral and legal reactions to misbehavior. As Angela Smith writes: "Blaming is a way of responding to faults in

---

62 Note that consequentialists can intervene already at this point and put forward a revised explanation of their own. They could argue that such driving is already part of being a bad exemplar, and this, which invites to public criticism and resentment aiming at reforming that person to behave better in the future. Consequentialism has to be regarded as a theory that suggests evaluating reactions to immoral behaviour on the same scale as the immoral behaviour itself (Harman 1977, 159–162). I have expressed a very general criticism about such convictions in section 4.7.

ourselves and others, and can be unfair or inappropriate [or fair and appropriate, on the contrary] for any number of reasons." (Smith 2007, 472) Rectification is amongst this number of reasons. Our legal system also follows the abstract idea closer to rectification as "restoring justice." Hence, putting the unlucky driver in jail is a reaction that should equalize (in an abstract manner) the death of the pedestrians, such as buying a new vase is a rectification of (even accidently) dropping one. This seems to be an inadequate response since taking a life is a damage that cannot be restored. However, the initial idea goes in this direction, and countries that still employ capital punishment are even closer to this "restorative mentality" of legal punishment. It is important to emphasize that I do not aim to justify an idea of "restorative justice." It should merely be mentioned that this is *in fact* one of the underlying, implicit ideas of many legal systems, and this idea can be applied only to unlucky drivers—when damage occurred. Moreover, legal punishment also serves other purposes that are rather detached from actual wrongdoing, namely deterrence and rehabilitation.[63] For people who hurt someone else (even out of sheer misfortune) are punished to *illustrate* that such misconduct—the acceptance of existential risks—is a serious disregard of our social and legal rules and is deemed unacceptable (Rachels 1993, 132). His punishment serves as a reminder of the validity of these statutes. Through the idea of rehabilitation, on the other hand, we understand the legal response to the unlucky driver not as a punishment, but as an opportunity to reflect on one's misconduct, change one's behavior and re-integrate them into society. The occurrence of damage is merely taken as an indicator and not as a foundation, that time for such a measure is suitable. The damage that occurred merely provides a date with which to start. It does

---

63   As I have pointed out in the previous chapter: Consequentialists will generally detach wrongdoing and blameworthiness by considering blaming as an action that is appropriate when it maximizes the good, and when it has a factual influence on the agent (Harman 1977, 160–161). But as much as in other cases of action and despite their factual importance, I do not believe that such motivation exhausts the varieties of values that play a role in practices like punishing and blaming. Clearly, the dispute about the appropriateness of restorative justice and deterrence in regard to fairness and control is then reproduced on this level and our ordinary convictions can (and should be) again directed against such detachment in principle (Rachels 1993, 115–116).

not substantiate the belief in the severity of his misconduct. Simultaneously, we must not employ diverging views on the lucky and unlucky driver's moral standing.

Again, I do not intend to justify that any or all those purposes conjoined should constitute the general way of dealing with legal offences. The previous reasoning should also not function as a justification for the differences in the evaluations of people who have been lucky enough to spare damage to others as well as those who were not equally fortunate. The above outlined aspects merely provide an *explanation* for the differences in our initial moral reactions. In our initial moral reactions, we mix up these diverging purposes and their different treatment of the lucky and the unlucky driver. When we separate them more clearly, we understand why it is appropriate to find both drivers both equally and morally responsible, while at the same time we can understand that we do treat them legally (and sometimes also morally) different. This is at least one way of responding to consequential moral luck that reduces the air of "paradox," or "inconsistency" that is associated with our diverging treatment of lucky and unlucky people.[64] Let us consider other instances of moral luck that are usually more closely associated with the triangle responsibility, determinism, and control.

5.2 Constitutive Moral Luck

Another kind of moral luck discussed by Nagel is worth of thorough consideration. At the beginning of the previous chapter I argued that moral responsibility is closely tied to the idea that decision are "up to the agent," as Robert Kane has

---

64   It should be mentioned that Thomas Schmidt is not convinced of this argumentation and finds it "difficult to see, how all of our relevant reactions could be interpreted as being grounded in something other than judgements about responsibility and blameworthiness." (Schmidt 2013, 304) His dissatisfaction is concerned with a form of reinterpreting moral luck proposed by David Enoch and Andrei Marmor. Schmidt understands Nagel's argument as being concerned with questions of blameworthiness, while the arguments presented above emphasize a distinction between being responsible and being held responsible (as being blamed) and I believe that a neglect of this distinction underlies the moral luck problem.

written (Kane 2007, 2). In a similar vein, Nagel argues that we are judging *persons* when we attribute moral responsibility rather than the arbitrary surroundings and "origins" of their actions or the fortunate consequences thereof (Nagel 1991b, 25). Persons that ought to be regarded as potential addressees of responsibility ascriptions must be understood as the authors of actions. Thus, we implicitly assert that they could have done something else than what they did or nothing at all. Let us consider the following case to make this intelligible. Usually we assume that someone, such as the famous German football manager Uli Hoeneß, who evaded taxes, is legally and morally responsible for what he did. This real life example resembles John Martin Fischer's Sam-case (he utilized it to comfort our compatibilist intuitions) although, we should not be concerned with the facts regarding the case for our current purpose. This example can be easily transferred to people working in innovation contexts such as visioneers and innovators whose responsibilities were discussed in chapter three. Nagel emphasizes the contingent and contextual conditions in which humans grow up that supposedly affect their opportunities and skills (Nagel 1991a). Nagel sees a normative problem arising from insights in the field of moral psychology.

Coming back to our examples: Hoeneß' misconduct was considered obviously illegal and of such quality that a prison sentence seemed adequate. However, skepticism concerning this conviction and our initial reproaches and judgment of his responsibility can easily arise if we think about the guilt of such a person from a very different angel. We might ask: What made him evade taxes? When we start speculating about the reasons he might have had, or consider that his action is just a particular instance that reflects certain longer lasting character traits which he might have never consciously adopted. It could be suggested, for example—based on our common sense intuitions—that he is a parsimonious person, thereby, disposed to break the law (or at least certain moral constraints) to foster financial profit. We might suggest that gaining financial profit seems to be an important goal for Uli Hoeneß, and he is willing to take measures that are beyond the realm of what is legal and do things that are at least morally questionable. Roughly speaking, we could consider this as expressions of his parsimoniousness.

But where does this character trait of parsimoniousness come from? What is its origin? How did Uli Hoeneß become the person he is? When we consider his personal development until the moment of his reprehensible action of tax evasion from an external standpoint, we might be inclined to regard his character less as a product of his own efforts as we initially did. Being born and raised in a catholic household in Southern Germany shortly after World War II this—so run our intuitions—must have affected him to being disposed in certain ways. We are inclined to believe that certain biographical dates, such as the social and cultural environment of his childhood, and large parts of his educational history had a substantial influence on his current character and actions (Burnyeat 1980, 70). All of those aspects were clearly out of his control. For him, as for everyone else, the circumstances of one's childhood are a matter of sheer luck, and they might have a major impact on one's future behaviour. In this perspective, he no longer seems responsible for his actions. Uli Hoeneß' current moral flaws seem to be the products of causes beyond his control. This is how Nagel characterizes our evaluative system of temperaments and other aspects of our human characters, which are beyond one's control. Whether you are born as a choleric person or as an introvert seems to be a matter of luck, and those factors have a major impact on our actions. Nagel writes:

> An envious person hates the greater success of others. He can be morally condemned as envious even if he congratulates them cordially and does nothing to denigrate or spoil their success. Conceit, likewise, need not be displayed. It is fully present in someone who cannot help dwelling with secret satisfaction on the superiority of his own achievements, talents, beauty, intelligence, or virtue. To some extent such a quality may be the product of earlier choices; to some extent it may be amenable to change by current actions. **But it is largely a matter of constitutive bad fortune** [own emphasis]. Yet people are morally condemned for such qualities, and esteemed for others equally beyond control of their will: they are assessed for what they are *like*. (Nagel 1991b, 33)

Although clearly reviewed, this passage sheds a different light on the previously discussed example. Nagel's interpretation is more moderate than the one presented above. According to Nagel, Hoeneß upbringing and temperamental predispositions do not "determine" his actions in a strong sense. Nagel has said previously that "[a] person may be greedy, envious, cowardly, cold, ungenerous, unkind, vain, or conceited, but *behave* [sic] perfectly by a monumental effort of will. To possess these vices is to be unable to help having certain feelings under certain circumstances, and to have strong spontaneous impulses to act badly." (Nagel 1991b, 32–

33) What Nagel apparently asserts is that our moral evaluations are sometimes directed at someone's behaviour but often at his or her character and, furthermore, that it is this character that people cannot help to have. A stronger explanation of Nagel's arguments has been written by Bruce Waller (Waller 2011). Waller argues similarly, as I tentatively did in the passage above, that people's unchosen character traits "determine" their actions and, henceforth, their responsibility. Both readings entail certain shortcomings that should be outlined in the following.

Let us first start by discussing Nagel's weaker claim, which consists apparently of two distinct assumptions: We assess people not only by their distinct actions but also by their wider character *and* this character is not within one's control. Nagel opposes the first notion to Kant's way of approaching moral goodness. Nagel emphasizes his opposition with the passage of Kant's beginning of the *Foundations of the Metaphysics of Morals* where Kant says that the free will alone is an appropriate object of praise and blame and, furthermore, that duties alone are within the reach of the will as determinable objects (Kant 1996, 18). Character traits are not amongst the possible objects of determination. Yet, Nagel argues, that Kant's conclusions are intuitively unacceptable (Nagel 1991b, 33). He holds, as in the above-cited passage, that we do indeed judge people for their feelings, temperaments and character traits, even if they behave perfectly according to, for instance deontological moral standards.

The present discussion is again located at the intersection between normative ethics and theoretical philosophy. These fields strongly conflate in discussions about responsibility. As will be outlined in more detail in chapter seven, there are compelling reasons to assess people such as visioneers and innovators in general primarily (not solely) regarding their character traits. Amongst the reasons to favor *aretaic* ethics is *in nuce* that the success of reductive theories usually clustered together as modern moral theories, such as deontology and utilitarianism, which primarily aim at evaluating people's conduct (separated usually into distinct actions). This is impaired by the fragmentation of values including those that are neither self-interested nor impartial by nature and do not fit in those modern ethical schemes (Nagel 1991d). They miss out relevant normative features, for example living a meaningful life (Wolf 2010). Therefore, it makes sense to focus on the layer of virtues and vices when evaluating people and their behavior (see chapter seven). Since Nagel starts his debate from the standpoint of ordinary morality, he is aware of the conflict that constitutive moral luck produces. Without judging whether his own critique of Kant affects its persuasiveness, I can conclude that I am sympathetic with Nagel's first assertions namely that virtues play an important role in our moral evaluations. Greed, envy, and parsimony are reprehensible character traits. But what should we think of the second assumption? Nagel himself is utterly imprecise when he talks about the nature of virtues. In some passages, he expresses that virtues are to some extend under one's control. In the above cited passage he says "[t]o some extent such a quality may be the product of earlier choices; to some extent it may be amenable to change by current actions." In another passage prior to this one, he argues more pessimistically that "[to] possess these vices is to be unable to help having certain feelings under certain circumstances, and to have strong spontaneous impulses to act badly." In this vagueness about the amenability of virtues lies the main problem of his argument about constitutive moral luck. Constitutive moral luck is a matter of fortune only if we really cannot help to be the persons we are. This seems to be the case for some of the things Nagel mentions and considers as virtues, such as emotions and impulses. However, in most virtue ethical theories these aspects of human psychology are not part of the scheme of virtue ethics. Impulses are not within the scope of voluntary determination (which might be analytically true): That is why we are not held responsible for them, as we are also not held responsible for crimes of passion such as acts of self-defense (Slote 2010, 140). In contrast: For character traits, such as greediness, envy, and others, this is, however, not a priori the case (Smith 2005). At least Nagel does not provide a reason for virtues not being an object of voluntary decision. Thus, in a less well-meaning interpretation, he merely *presupposes* non-amiability without reason. In a more well meaning interpretation he rather *denies* non-amiability, which means, however, that constitutive moral luck is not as threatening as it seemed to be. The stronger interpretation of his reasoning is supported by an interpretation of Aristotle that also Judith Andre employs when she writes:

> Part of being moral, for most of us, is being virtuous; and being virtuous involves more than doing the right thing. It involves as well the ability to see what the right thing to do is, and the desire to do that right thing. […] Virtues, as Aristotle describes them, are possible only to those who have been reared in a moral community; a fortunate childhood fosters adults who feel rightly as well as acting rightly.

> But people cannot choose their own upbringing, and emotions are involuntary.
> (Andre 1983, 204)

This coincides with the strong interpretation of Nagel's reasoning and—as indicated above—at least the first part of this passage resonates well with my own point of view. However, also Judith Andre *softens* the point made in the second half of the passage that concerns constitutive moral luck when she writes (in brackets) that "[w]e do have indirect control over our emotions [...] but this is limited." (ibid., p. 204) It must be emphasized that Judith Andre's Aristotle interpretation is not in line with the understanding of virtue ethics that will be presented in chapter seven of the present book. There are many passages in Aristotle's *Nicomachean Ethics* that suggest that virtues are at least indirectly under one's control when properly trained (Burnyeat 1980, 73). In Passage 1103a 14–27, for instance, he argues that one cannot "train fire to burn downwards," and on the contrary: The full development of the moral virtues "in us is due to habit." Also James Urmson writes in his interpretation of Aristotle's ethics: "If properly **trained** one comes to enjoy doing things the right way, to want to do things the right way, and to be distressed by doing things wrongly [own emphasis]." (Urmson 1999, 26) And later:

> Aristotle compares acquiring a good character with acquiring a skill. [...] Before one has acquired the art or skill one acts in accordance with the instructions of a teacher, who tells us what to do, and one does it with effort. Gradually, by practice and repetition, it becomes effortless and second nature. (Urmson 1999, 26)

The foundations of this interpretation are the passages in the *Nicomachean Ethics* in which Aristotle compares virtuous behaviour with mastering an instrument (1179b 20–26). Clearly, Aristotle was aware of the influence of upbringing and the social and political environment for the development of one's character (1103a 19–20). However, he was not as pessimistic as Andre and Nagel (in the strong interpretation) were about the possibility of shaping these traits. We have an *indirect* control over our behavioral patterns and dispositions, which is why constitutional moral luck is a problem only on the surface (Pauen 2004, 101). We are, therefore, coherent in judging people's psychological patterns because there *is* a substantial degree of influence that they can exercise over them. A virtuous person will, for instance figure out ways of dealing with her temper (Smith 2005; Bieri 2015, 76). Accordingly, a first step could be to recognize which aspects of one's surroundings trigger certain reactions or feelings. If one is disposed to be easily upset when in places where many people gather, or when being under the influence of alcohol, it might be helpful to avoid these contexts. Later on, one might realize that the actual origin of one's arousal was a different one that also occurred in those contexts. Such realization can lead to make the temper vanish completely. One does not directly chose to drop such dispositions, but one chooses (or not chooses) proper training methods and strategies to handle them (if they are not admirable), and this can eventually lead to getting rid of them. It would be surprising, in Geert Keil's words, if everyone would experience the same obstacles and (internal) resistances in meeting moral demands (Keil 2007, 165).

As mentioned before, another interpretation is even stronger than our previous strong interpretation of Nagel's argument. Nagel argues (in the strong interpretation) that people can ("by a monumental effort of will") control their behavior, but we judge them for having the underlying emotions and dispositions. It is these traits, which they naturally possess. Bruce Waller takes up Nagel's thoughts and presents an argument in which—as in Nagel's case—normative ethics and the general problem of finding conditions for being held responsible are intertwined. The first suggestion Waller makes is that constitutive moral luck has a much stronger influence on our life's courses than we usually think. It seems that being a person who is disposed in a particular way also predetermines that person's life-course and one cannot help it. Waller's second suggestion seems to be that we should drop the notion of moral responsibility altogether because of our lack of control over these initial dispositions. To illustrate these points Waller utilizes an example of Alfred Mele. Mele asks us to imaging a child, which he calls Betty. She is six years old and afraid of the basement. Betty realizes that her fear of the basement is "childish," and she decides to overcome it. She periodically visits the basement until she overcomes her fear. Mele suggests that if she succeeds in getting rid of her fear through her strategy of facing it, this can be considered as an act of "intentional self-modification." (Waller 2011, 29) In contrast, Waller places alongside Betty a fictitious twin brother Benji, who is, according to Waller, "not quite as brave as his twin sister." For this reason, he does not successfully enter the basement, and relies on his mother instead and, thus, develops a psychological pattern that Waller calls "cognitive miserliness" which is constitutive for Benji's later behavior. Waller writes:

> Place alongside Betty her six-year-old twin brother, Benji, who also suffers from fear of his basement (and who, like Betty, knows that no harm has befallen those who venture there). Benji also regards this fear as "childish" and wishes to get beyond it. But Benji is a little—just a little—less self-confident than his sister. Rather than taking bold steps to deal with his fear, Benji decides to wait it out: maybe I'll grow a bit bolder as I grow older, Benji thinks; besides, Mom is plenty strong and courageous, so there's no need for me to make an effort that might well fail. Betty has thought up a good plan, Benji recognizes, but well-planned projects often come to a bad end, like that well-thought-out plan to stand on a chair to reach the cookie jar. Benji is not quite as strong as his sister, in some very significant respects. He does not have her high level of self-confidence (or sense of self-efficacy); his sister has a strong internal locus-of-control, but Benji is inclined to see the locus-of-control residing in powerful others. And although Betty is well on her way to becoming a chronic cognizer (to be discussed shortly), Benji has developed significant tendencies toward cognitive miserliness (the abysmal failure of that well-thought-out campaign to liberate the cookie jar left a deep mark); that is, even at this tender age, **Betty and Benji already have significant differences (not of their own making or choosing)** [own emphasis]. (Waller 2011, 31)

To make the point with a regard to responsibility even stronger, he argues that those initial traits and characteristics are constitutive for the differences in one's later behavior, and it would be unfair to base moral evaluations on aspects of one's constitution that one has not chosen and could not have affected:

> [...] we are each different in our capacities and talents and cognitive abilities and fortitude; careful comparisons of those differences in character and history soon undercut any claims or ascriptions of moral responsibility. Those differences make it unfair to blame one and reward another for their **differences in behavior** [own emphasis]. (Waller 2011, 40)

As I mentioned before, Nagel assumes that people can overcome their dispositions by monumental efforts of will, but they cannot help having these dispositions. In Waller's description, the opposite seems to be the case. People have certain dispositions, which are constitutive for their resulting behaviour. The first crucial question is: How should we understand the term "decide" in the first sentence marked

boldly? Waller writes that "Benji decides to wait out." It seems—if we also consider the second quote—that Waller argues that Benji could *not* have chosen to act
like his sister because of his general fearfulness and lack of confidence. His
slightly greater lack of confidence is the source of his tendency to avoid. If the
passage is understood in this sense, Waller invests in his case description an interpretation of the concept of a "disposition" that is much stronger than Nagel's. It
suggests that Benji could not have done anything to master his fear successfully.
It seemed impossible to him to employ the same strategy to overcome fear as his
sister. If this is Waller's suggestion here, than we should not be surprised that this
undermines his responsibility. Benji is described as a "slave of his passion" rather
than as a rational agent (if we are allowed for the sake of argument to attribute this
term to a six year old boy). He lacks the kind of control that we presumed to be
necessary for responsible agency. However, that also means that with this conceptual investment Waller begs the question of Benji's responsibility. In this strong
interpretation, Waller conceives Benji so that he could not have done anything
different. This puts Benji on the same level as people with psychological illnesses,
such as an obsessive disorder. Then again, whether his fear really has this quality
is an open question, such as in normal real life analogues. Campbell lucidly points
out that the very notion of acting implies the decision against one's desires or pre-
figurate inclinations and to act, for instance according to moral reason: "But all
that that amounts to is that formed character prescribes the nature of the situation
within which the act of moral decision takes place. [...]. For the very nature of that
decision, as it presents itself to him, is as to whether he will or will not permit his
formed character to dictate his action." (Campbell 1967a, 43)[65] If on the other hand
we understand Benji's decision-making process in the way outlined by Waller as
the proper reflection of different options about the necessity to overcome his fear

---

[65] The argument presented here can also be directed against Galen Strawson's reasoning in his Imposssibility of Moral Responsibility. My suggestion is that the same
petitio is at place when Strawson asserts as a first premise of his argument: "(1)
You do what you do because of the way you are." (Strawson 2013, 315) The right
description, in contrast, would be: You do what you do because you find this or that
reason more convincing. One does not have to be a causa sui to be an agent as
Strawson suggests. One also does not always have to have reasons for action, but
sometimes.

and the possibility to avoid going to the basement, there is reason to consider this as a decision for which he is fully responsible—if we can at all attribute this notion to a six year old boy—and for which we can imagine he will also stand up in the future (viz. accept his responsibility). We can imagine, for instance that his mom will ask one day why he never goes to the basement, and he outlines exactly these reasons: "You know, I do not feel well going down there and as long as you pick up the potatoes, I also do not have to. Is that alright for you?" In this respect, fearfulness has been a parameter in his decision not to test himself and accept staying away from the basement, as much as the insight that there is no urge for him to go down. Fearfulness has not *determined* this decision and this is all that is needed for responsible agency. The case is somewhat problematic since both Benji and Betty cannot be considered as full-fledged responsible agents (because of their young age) and avoiding the basement does not have any (moral) significance, although with some of these choices both of them lay (amiable) foundations for other choices later in life. But consider that Benji will face many analogous situations later in life. Imagine him, for instance, being twenty years old and having a much wider perspective, increased knowledge and a wider and more comprehensive set of reasons to balance. He might still be fearful and wrestle with himself occasionally, for instance when he has to get a driver's license. He is afraid of being involved in an accident and despises the idea of being responsible for the lives of others as well as expensive vehicles. He is also aware of all the side effects such hesitation has, for instance that this might limit his mobility and job opportunities. We expect him to face his demons or to arrange things in a way that makes living a meaningful and ethical life possible. There are plenty of possibilities for living without a driver's license; he could move to a cycling friendly community or become more accessible through modern means of telecommunication. If he wanted to, however, become an emergency medical assistant or a police officer, or if his parents are sick and live on the countryside where transportation without a car is more difficult, he is forced to find a different solution. He might consider becoming adjusted to traffic by first getting a moped license (which means riding a less dangerous vehicle), or driving more often as a front-passenger before taking over the steer of a car. He might go more often to amusement parks and get adjusted to the high speeds of rollercoasters or go-carts or do whatever helps him to control his anxiety (Smith 2005, 253). If it turns out that his fear has a chronic or compulsive character, we expect him to search for professional advice (Foot 1978, 12). We demand from Benji as from any other "normal" adult person that he employs strategies of this sort if there are compelling (moral or prudential) reasons to do so (Seebaß 1993a, 227). These are amongst the normative standards we set for living a responsible life.

The present section does not intend to substitute a guidebook for practical decision-making. It should be sufficiently clear why constitutive moral luck does not undermine responsibility. There is indirect control over our character traits and dispositions, and we are responsible to train, ameliorate or dispose of them if necessary.

5.3   Perspectives and Incompatibility

As mentioned above, Nagel has an overall theme, which he refers to after his discussion of moral luck. This broader theme concerns different perspectives on human action, and it is suggested to baffle our understanding and trust in responsibility and autonomy. His point about the incommensurability of the external and the internal perspective rests on the arguments for constitutive and consequential moral luck as much on the later discussed explanatory problem. Although, it can also be discussed entirely separate from these examples. Nagel says that the "desire for a comprehensive picture of objective reality" is one of the strongest motives in philosophy (Nagel 1986, 13). Since Nagel understands in several passages of VFN the inclination to view ourselves from the external perspective as something "natural" or inevitable, we seem to be inclined to adopt an external viewpoint on ourselves with or without the additional arguments of the explanatory gap which will be discussed later on or instances of moral luck which were analyzised before. Against Peter Strawson's view of the "inescapability" of our reactive attitudes and the moral sentiments (Strawson 1985, 33)—a position that will also be considered in more detail later on—Nagel claims that "there is no way of preventing to slide from internal to external criticism once we are capable of an external view." (Nagel 1986, 125) This is why I believe one can treat the perspectivity-argument—as one might call it—separately. I will provide two objections against Nagel's conception of the external perspective as being incompatible with our subjective point of view, and both are related to each other.

The first point aims at understanding the argument from externality thoroughly and suggests—based on this understanding—that we are easily capable of switching between different perspectives and standpoints including the external; our own and those of other people. This understanding will allow me to transition to the second point, which says that Nagel's claim of the incompatibility of these perspectives is not as significant and devastating as suggested. We can interpret the adoption of those perspectives as a tentative and often constructive method to extend different sorts of knowledge. Looking at the variety of viewpoints adoptable in this way does not force us into intellectual struggles or difficulties and often proves valuable to see whether things that appeared initially comprehensible (or not) are really that way. In ML, Nagel claims that the external point of view of the agent is essentially a view from which the world is mainly conceived as a series of events. He writes that "[t]he problem arises [...] because the self which acts and is the object of moral judgment is threatened with dissolution by the absorption of its acts and impulses into the class of events." (Nagel 1991b, 39) In VFN he makes this claim as well but puts it on a more general level. Here he argues that insecurity and doubt about autonomy and responsibility can result from adopting the idea that the world is determined, but already the event-ontological perspective on the world leads to a dissolution of agency and to distance from the idea of the agent's active engagement with the world. The opposition between the internal and the external perspective in VFN is described as follows:

> While we cannot fully occupy this perspective towards ourselves while acting, it seems possible that many of the alternatives that appear to lie open when viewed from an internal perspective would seem closed from this outer point of view, if we could take it up. And even if some of them are left open, given a complete specification of the condition of the agent and the circumstances of action, it is not clear how this would leave anything further for the agent to contribute to the outcome anything that he could contribute as source, rather than merely as the scene of the outcome the person whose act it is. If they are left open given everything about him, what does he have to do with the result? From an external perspective, then, the agent and everything about him seems to be swallowed up by the circumstances. This happens whether or not the relation between action and its antecedent conditions is conceived as deterministic. In either case we cease to

face the world and instead become parts of it; we and our lives are seen as products and manifestations of the world as a whole. (Nagel 1986, 113)[66]

This incommensurability and the rise of insecurity described here with reference to one's own autonomy have its analogue in the realm of responsibility, which is concerned with the autonomy of others. With dissolving the agent and the alternatives that lie open before him in a world entirely composed of events that are connected as causes and effects (Nagel speaks of necessitation "by prior conditions and events" (ibid., p. 115)) the appropriateness of our moral sentiments and reactive attitudes seem to be undermined and we lose our trust in moral responsibility. Nagel writes:

> The same external view that poses a threat to my own autonomy also threatens my sense of the autonomy of others, and this in turn makes them come to seem inappropriate objects of admiration and contempt, resentment and gratitude, blame and praise. (Nagel 1986, 112)

And also:

> The main thing we do [when assessing someone's responsibility] is to compare the act or motivation with alternatives, better or worse, which were deliberately or implicitly rejected though their acceptance in the circumstances would have been motivationally comprehensible. That is the setting into which one projects both an internal understanding of the action and a judgment of what should have been done. It is the sense of the act in contrast with alternatives not taken, together with a normative assessment of those alternatives also projected into the point of

---

66   It is worth mentioning that Charles Campbell establishes the opposition between the internal and the external perspective regarding action similarly in an Inaugural Lecture in 1938 which is reprinted in (Campbell 1967a, 48): "Now this criticism […] seem to me to be the product of one simple, but extraordinarily pervasive error: the error of confining one's self to the categories of the external observer in dealing with the actions of human agents. […] It is perfectly true that the standpoint of the external observer, which we are obliged to adopt in dealing with physical processes, does not furnish us with even a glimmering of a notion of what can be meant by an entity which acts causally […]. But then we are not obliged to confine ourselves to external observation in dealing with the human agent." See also his The Psychology of Effort of the Will (Campbell 1967b, 75).

> view of the defendant that yields an internal judgment of responsibility. What was done is seen as a selection by the defendant from the array of possibilities with which he was faced, and is defined by contrast with those possibilities. When we hold the defendant responsible, the result is not merely a description of his character, but a vicarious occupation of his point of view and evaluation of his action from within it. (Nagel 1986, 121)

Ordinarily when we seek to assess someone's responsibility, we try to understand his behaviour from his own perspective. Through this move of self-transcendence, we already leave our subjective point of view in order to understand the other person's behavior. We put ourselves regularly "in the shoes of other person." This "ordinary" transcendence of the subjective perspective does not result in the sort of unease or bafflement that Nagel closely connects to the external point of view. However, we can also apply this move of self-transcendence and transition of perspectives in a number of other situations and contexts.[67] Imagine you are watching an exciting boxing match. You are not a full-hearted fan of either of the two fighters. After twelve rounds of relentless fighting, we imagine going to the scorecards and one of them winning a majority decision. The winner cheers to the crowd proudly and satisfied over his win. He has defended his championship belt. Then, the camera switches to the looser of the bout; he fought thirty-six minutes in relentless battle, was injured and finally beaten. Sad and devastated, he leaves the ring without comment. It does not require much empathy to understand and feel both the satisfaction and happiness of the victor as much as the sadness of the looser. These standpoints seem to be at first sight as incompatible as Nagel's external and internal viewpoint regarding action. Still, we can adopt both positions and understand the diverging feelings they carry at the same time. The transition between theses perspectives is often extremely smooth and unnoticed. One might even argue that the excitement of such fights stems from our ability to transition permanently between the two fighters' perspectives and understand their suffering as well as what winning and losing means to them (which presumes the winning and losing of the other). We do not have to decide on who's side we are standing, but can perfectly maintain the incompatible perspectives of both of them (which

---

67   Richard Wollheim describes in detail a number of cases inspired by psychoanalysis (Wollheim 1999, 64–84).

are not our own) while watching. What I am arguing here is that the subjective and the objective point of view are incompatible. They are somehow incompatible just as the two fighters' viewpoints. However, what is also emphasized by this example is that this collision of viewpoints does not necessarily result in the kind of unease or bafflement that Nagel assumes to befalls us. Nagel argues in VFN that the problems of autonomy and responsibility are not primarily verbal. He does advocate the view that recognizing the incompatibility of the external and the internal view on actions results in a fundamental "disturbance of the spirit," in "unease," in "affective detachment," and "imbalance." (ibid., p. 112) In contrast, what I showed through my example is that we can regard transitions in perspectives in a different light and, thereby, take the sting as much as the bafflement that might be accompanied away. Let me briefly contrast this viewpoint with Peter Strawson's critique of Nagel's perspective-argument.

Strawson essentially agrees with Nagel's reasoning regarding the opposition between the external—in Strawson's words "purely objective" or "purely naturalistic"—and the internal point of view which we "naturally occupy as social beings." (Strawson 1985, 35) He adds, however, two objections. First, he outlines with reference to his own view presented in *Freedom and Resentment* (FR) that we cannot take the external position "for more than a limited period in limited connections." (Strawson 1985, 36) I will return to this position and discuss it in more detail in the next section. The second objection Strawson makes concerns the idea that one has to force a decision between the subjective and the objective point of view, which seems to have implied such reasoning.[68] As a source of unease and given their lack of compatibility, we must choose between them:

> What I want now to suggest is that error lies not one side or the other of these two contrasting positions, but in the attempt to force the choice between them. The question was: From which standpoint do we see things as they really are? and it carried the implication that the answer cannot be: from both. It is this implication that I want to dispute. But surely, it may be said, two contradictory views cannot be true; it cannot be the case *both* that there really is such a thing as moral desert

---

68   Nagel is more phlegmatic about the possibility of choosing between them. I think that he believed that they are incompatible, but also equally inescapable. He describes a sort of stalemate.

and that there is no such thing, *both* that some human actions really are morally praiseworthy or blameworthy *and* that no actions have these properties. I want to say that the appearance of contradiction arises only if we assume the existence of some metaphysically absolute standpoint from which we can judge between the two standpoints I have been contrasting. But there is no such superior standpoint […]. (Strawson 1985, 37)

In FR he calls the subjective point of view the "participant's view" and also argues that the viewpoints are opposed to each other, but not exclusively (Strawson 1962, 9–10). I think that Strawson points in the right direction when he argues that we do not have to force a choice between them. Perspectives, even when they are incompatible can, as I argued before, coexist. My own reasoning, however, does not require Strawson's assumption of the lack of a metaphysical point of view from which we could judge which of the two perspectives is "more real." My suggestion is rather that we must not force the opposition between the two standpoints because they are not contradictory, but *merely incompatible*. In fact, by "de-verbalizing" the problem of autonomy and responsibility (in at least this part of his reasoning), Nagel substantially weakens the opposition between the external and the internal point of view. When we enter the realm of perspectives and viewpoints, we leave the realm of truth-functional statements—or in other words: the realm of *propositions*. Things then merely appear in certain ways. This is why Strawson phrases the problem wrong when writing that two "contradictory views cannot be true." When Nagel speaks about viewpoints, he does not conceive them as attitudes with propositional content. Believing that p and seeing p from such and such a perspective are entirely different things.

It would, therefore, be more appropriate to rephrase the opposition of viewpoints in contrast to Strawson as follows: on the one hand it *appears* or *seems* that there is such a thing as moral desert *and* that there is no such thing and, furthermore, that it appears or seems that human actions are really praiseworthy or blameworthy and that no actions have these properties. This is not the same as articulating these sentences suggesting that they have propositional content, such as when I say: "I believe that p and non-p," which is an apparent contradiction. Believing, that it *appears* as if p, and as if non-p is not equally contradictory. Due to this reason, we do not have to force a choice between the internal and external point of view. Internal and external perspectives properly understood as perspectives can even out when being incompatible coexist simultaneously. Moreover, when we consider that we use the transition of perspectives frequently without creating unease or bafflement as mentioned above, but rather as a method to acquire new ideas, understanding, and to advance our knowledge, we take the sting away from their opposition. When we, for instance, write a novel, we adopt fictitious people's perspectives to comprehend and set out what would be the most natural thing to do given a protagonist's character and history. The same we do with the anti-protagonist including her opposing goals and motives and with all other figures in the play. Our purpose is to create a realistic and dramatic constellation between these figures. Like in the area of literary art, the transitions between perspectives can be a fruitful methodology.[69] Approaching the transition between perspectives this way, we realize that their nature is less destructive as Nagel suggests when considering them as a source of bafflement, unease, and insecurity.

5.4 Moral Sentiments and Reactive Attitudes

We have discussed Nagel's arguments in extent and Strawson's responses to them. Strawson has proposed an original theory that is considered as a form of agnostic compatibilism (Keil 2007, 63). His position is considered agnostic because it leaves the question about the truth of determinism open. Strawson provocatively claims to not know what the thesis of determinism actually is (Strawson 1962, 1). The theory of moral sentiments had a number of notable advocates until recently, but also famous forerunners, which will be mentioned briefly (Wallace 1994). It is first advanced in the landmark article *Freedom and Resentment* from 1962, which is based on his British Academy lecture (Strawson 1962). While Strawson provides the theory of the moral sentiments as a sort of a compatibilist theory of responsibility, his arguments require a radically different response than the ones presented against pragmatist and other compatibilist theories in chapter four. The basic thesis of the theory of moral sentiments is that our reactive attitudes and feelings of "offended parties and beneficiaries [...] such things as gratitude, resentment, forgiveness, love, and hurt feelings" are essentially expressions of a *life*

---

69 Strawson mentions the understanding of opposing moral viewpoints by taking the stance of other agents (Strawson 1985, 46–47).

form which cannot be fully given up, even if determinism were true (Strawson 1962, 4). The moral sentiments and our reactive attitudes are an inescapable part of human nature. They can be refined by temporarily adopting a more "objective" point of view but not entirely reasoned-away through insight into the mechanisms of the universe.[70] It is this idea that has most relevance in the discussion about determinism and responsibility and about which many philosophers suggested that it has been the most influential aspect of Strawson's theory of moral sentiments (Wallace 1994, 11). Before I discuss Strawson's ideas critically, I will introduce them in the following order with a historical reference point. The main element of Strawson's theory has been pre-formulated by David Hume in his ethical writings. This is astonishing insofar as Strawson never refers to these works of Hume. Consider the following passage from Hume's *Principles of Morals*:

> But when Nero killed Agrippina, all the relations between himself and the person, and all the circumstances of the fact, were previously known to him; but the motive of revenge, or fear, or interest, prevailed in his savage heart over the sentiments of duty and humanity. And when we express that detestation against him to which he himself, in a little time, become insensible, it is not that we see any relations, pf which he was ignorant; but that, from the rectitude of our disposition, we feel sentiments against which he was hardened from flattery and along perseverance in the most enormous crimes. In these sentiments then, not in a discovery of relations of any kind, do all moral determinations consist. Before we can pretend to form any decision of this kind, everything must be known and ascertained on the side of the object or action. **Nothing remains but to feel, on our part, some sentiment of blame or approbation; whence we pronounce the action criminal or virtuous** [own emphasis]. (Hume 1975, 290–291)

In a following passage of the *Enquiry Concerning Human Understanding*, Hume takes the existence of the moral sentiments as the explanation for our diverging reactions towards the actions of humans and inanimate objects such as trees: "inanimate objects may bear to each other all the same relations which we observe in moral agents; though the former can never be the object of lave or hatred, nor are

---

70   Nagel thought (and argued against Strawson) that the transition to a more objective point of view is inescapable (Nagel 1986, 125). We see that Strawson does not deny this.

consequently susceptible of merit or iniquity." (ibid., p. 293) In Hume's work, this is where the difference between practices of domestication and Paul Russel points out this striking similarity between Strawson's and Hume's conception of the moral sentiments and emphasizes the crucial role they play in the theories of both authors:

> The fundamental point [Hume and Strawson] agree about is that we cannot understand the nature and conditions of moral responsibility without reference to the crucial role that moral sentiment plays in this sphere. (Russell 2008)[71]

While the reference to Hume's thinking is reduced in FR to a footnote in which he refers to Hume's way of reconciling of skepticism and naturalism regarding the nature of induction, he makes the connection in his later writing, *Skepticism and Naturalism: Some Varieties*, from 1985 more explicit. Although, Strawson also does not refer here to Hume's moral philosophy but rather only to his ideas about the problems of induction set out in his *Treatise of Human Nature*. Thus, it might seem as if Strawson was not aware of the theory of the moral sentiments outlined in the *Principle of Morals*, from which I cited before. Either that or Strawson decided consciously against such reference in virtue of the fact that Hume, in contrast to him, never explicitly discussed (or even asserted) the impossibility of disposing the moral sentiments altogether in his writings on moral philosophy. Hume—as Paul Russell rightly points out—merely emphasizes their fundamental role in evaluating different types of events, such as voluntary actions and inanimate behavior, and we know that Strawson goes beyond merely asserting such functional roles of the moral sentiments. In any case, it is Hume's assertion about the inescapability of the "habit of induction," the adoption of inductive principles that cannot be shaken by a fundamental skepticism on rational grounds that Strawson adopts and transfers to the problem of freedom and responsibility. Hume's naturalist idea in the realm of epistemology is rephrased by Strawson as "the inescapable natural commitment [...] to a general frame of belief-formation and to a general style (the inductive) of belief-formation. But within that frame and style, the requirement of Reason, that our beliefs should form a consistent and coherent system, may be

---

71   See also (Russell 1990).

given full play." (ibid., p. 14) Let us consider how these ideas are transferred to the discussion on freedom and responsibility.

Strawson outlines his basic idea by analogizing the general theme of the moral sentiments with reference to Hume's idea about radical skepticism concerning the external world and the nature of induction. He starts his outline by stating his doubts about the intelligibility of the libertarian notion of free will. This assertion already appears in the end of his *Freedom and Resentment* as the remark about "the obscure and panicky metaphysics of libertarianism." (p. 24) This is not essential to his argument, yet it might indicate an inexplicit underlying motivation for his promotion of a compatibilist theory (Keil 2007, 64). The doubt in the intelligibility of genuine freedom is what Strawson shares with many compatibilists, such as Hume (Hume 1975, 96), Schlick and others (Schlick 1962c, 157–158; Mackie 1990, 226; Wallace 1994). For these thinkers the idea of genuine free will is neither intelligible nor necessary for being responsible. Usually then, as mentioned before, genuine free will is understood as a kind of causal power that stands somehow outside the regular order of nature that presumes one could have acted differently given exactly the same conditions. This is suggested to be a meaningless idea lifting the decisions and actions on a par with random events. I will discuss this problem in more detail in the next section. In his book *Skepticism and Naturalism*, Strawson writes:

> […] for no one has been able to state intelligibly what such a condition of freedom, supposed to be necessary to ground our moral attitudes and judgements, would actually consist in. Such attempts at counter-argument are misguided; and not merely because they are unsuccessful or unintelligible. They are misguided also for the reasons for which counter-arguments to other forms of skepticism have been seen to be misguided; i.e. because the arguments they are directed at are totally inefficacious. We can no more be reasoned out of our proneness to personal and moral reactive attitudes in general than we can be reasoned out of our belief in the existence of body. […] our general proneness to these attitudes and reactions is inextricably bound up with that involvement in personal and social interrelationships which begins with our lives, which develops and complicates itself in a great variety of ways throughout our lives and which is, one might say, a condition of our humanity. What we have, in our inescapable commitment to these attitudes and feelings, is a natural fact, something as deeply rooted in our natures as our existence as social beings. (Strawson 1985, 32)

This is the central idea of Strawson's agnostic compatibilism. Social and personal relationships are a condition of humanity (of human nature), and they involve a kind of attachment to other people that entails moral sentiments and reactive attitudes. These attitudes and sentiments in particular cases can be refined, but the systems as such cannot be disposed even if determinism were true. This theory gives rise to a number of questions that are "internal" to the theory such as: When exactly do those moral sentiments occur (which is an empirical, descriptive question), and when are they appropriate (which is a normative question that involves normative ethical reasoning)? Strawson writes that there are occasions where we "shall not feel resentment against a man" when he has not been "himself" when acting or when he is a child, neurotic, or deranged (Strawson 1962, 9), which indicates a kind of control over the width and severity of those moral sentiments. For common purposes, we temporarily take a more objective stance and suppress our initial feelings, a stance that we cannot adopt for too long. When factual aspects about a person's behavior are revealed (such as the influence of a drug), we are able to modify our sentiments and our reactions. Also internal is the question to which degree the moral sentiments can be "modified or mollified," as Strawson says (which seem to be descriptive questions again) (Strawson 1962, 8), although regarding the overall aim of the theory we certainly *cannot* reply: completely abandoned.

Since Strawson's theory is a kind of compatibilism, which stands somehow queer to other compatibilist theories, I will focus on the external questions regarding this theory's general plausibility.[72] As a reminder: compatibilists believe that we have a kind of freedom (as long as we are not enchained or under external compulsion) that is not threatened if determinism turns out to be true. Strawson in contrast does not even speak about the relation of freedom and determinism. He

---

72   For a valuable and extended discussion of the internal aspects, see (Watson 2004c). Watson emphasizes that the original aspect in Strawson's approach lies in offering an alternative theory to the libertarian account of responsibility presuming a sort of "metaphysical person" and to what I called effect compatibilism that sees responsibility ascriptions as means of social regulation (Watson 2004c, 221). The most pressing internal questions seem to be how fundamentally those sentiments can be revised and according to which standards. Strawson's own account, rich as it is, is quite unsystematic.

asserts that our moral sentiments and reactive attitudes are elements of human nature that cannot be dropped by insight in determinism's truth or other rational arguments, and because of this, the very questions determinism's truth becomes *irrelevant* (Strawson 1962, 20). This is a very different *thesis* than that of traditional compatibilism; it equals the postulation of an *anthropological constant*. As Gary Watson paraphrases Strawson's position: "Holding responsible [in Strawsonian terms] is as natural and primitive in human life as friendship and animosity, sympathy and antipathy. It rests on needs and concerns that are not so much to be justified as acknowledged." (Watson 2004c, 22–223) We should ask at this point— similarly as we did about the thesis of determinism—: What makes such a thesis of the inescapability of the moral sentiments plausible, and what would it take to refute it? The following quote is the closest we get to a defense of Strawson's main point in his article:

> The human commitment to participation in ordinary inter-personal relationships is, I think, too thoroughgoing and deeply rooted for us to take seriously the thought that a general theoretical conviction might so change our world that, in it, there were no longer any such things as inter-personal relationships as we normally understand them; and being involved in inter-personal relationships as we normally understand them precisely is being exposed to the range of reactive attitudes and feelings that is in question. This, then, is a part of the reply to our question. A sustained objectivity of inter-personal attitude, and the human isolation which that would entail, does not seem to be something of which human beings would be capable, even if some general truth were a theoretical ground for it. (Strawson 1962, 12)[73]

---

73   There is another argument in FR that goes in short: If determinism were true, there would still be a difference between people who are exempt from moral responsibility (say insane people) and those we consider currently to be "normal," and only the compatibilist can uphold this distinction, for the incompatibilist believes that any difference between them vanishes (Strawson 1962). I mention in chapter four that incompatibilists do not have to deny that there is a difference between "normal" adults and insane people. What they deny is that the differences were still morally significant, if determinism were true. The compatibilist is forced to explain wherein the differences lie and will, thus, refer to suspicious notions of "control" or the ability to "act otherwise." I have addressed these objections in length at John Martin Fischer's approach in section 4.5.

Here, the reasoning is the same as presented above, which utilizes the notion of "nature" to indicate the depth of the moral sentiments. The depth or boldness of the moral sentiments has its analogue in our belief in the external world. Strawson connects the idea of the moral sentiments with our commitment to personal relationships. People involved in friendships and loving relationships cannot, as he claims, plainly adopt an impartial point of view. In many cases, this would undermine the special commitment and subsequent treatment of people involved in such relationships. However, can this be a ground for assuming that such objective viewpoints cannot be held for long or even forever, as Strawson claims? Strawson's argument is a shrewd way of sidestepping the actual problem of freedom and determinism and allegedly showing that two of the most pressing philosophical problems are irrelevant. The fundamental thesis on which it rests concerns the inescapability of the moral sentiments, and it is remarkable that Strawson does not actually provide evidence for this belief. It rests mainly on Strawson's assurance based on his inclinations. This is not only a thin basis for an argument it is also a *false* assumption when generalized. There are people who argue that one can (and should) give up holding anyone morally responsible (although we do not know whether these persons successfully "tamed" their sentiments, whether they are still attached to their personal relationships and so on) (Waller 2011).[74] Although, one must acknowledge that Strawson's assumption is hard to refute. The possibility that the entire scheme of reactive attitudes and moral sentiments cannot be relin-

---

74   Derk Pereboom argues that hard incompatibilists (who believe that genuine freedom is necessary for being responsible and that it is such freedom is an incoherent concept), such as himself, believe that some "attitudes have presuppositions that the hard incompatibilist would believe to be false." (Pereboom 2007, 119) Other moral sentiments and reactive attitudes are, however, unaffected by his stance. He writes: "However, suppose that you behave immorally, but because you endorse hard incompatibilism, you deny that you are blameworthy. Instead, you acknowledge that you have done wrong, you feel sad that you were the agent of wrongdoing, and you deeply regret what you have done [...]. None of this is jeopardized by hard incompatibilism." (p. 121) I am not convinced that this would be the resulting stance of someone who finds out that she is controlled by an alien lifeform. As argued before in the previous chapter such insight possibly causes severe detachment.

quished persists—which means on the other hand that is not necessary that determinism undermines some sense of moral responsibility. The reason is that the only occurrence that would unambiguously falsify such outlook necessarily lies in the future. Only, if everyone (or since there is always some deviation; a great number of people) abandons their moral sentiments and reactive attitudes entirely, Strawson's claim of the necessary connection between human nature and moral sentiments were falsified. Such a scenario is certainly *imaginable*.

Imagine, for instance, that I found out that my wife is a spy from another country, sent to get information from my employer, and she has been threatened to do the job by the life of her daughter (who I did not know before). I can imagine that she convinces me that she actually really loves me. She was compelled to be with me because the agency required information, but as soon as she started "doing her job" real feelings of love arose. I might think: "Why not? When I start playing a game initially intending to win some money and thus playing goal-oriented, I might nevertheless be roped in it after a while and enjoy the game purely for its own sake, forgetting about my initial aim. This "intrinsic pleasure" which is completely devoid of any instrumental value is nevertheless a *real* pleasure (Harman 1977, 149; Pereboom 2007, 121–122). Thus, I may conclude: "Love does not necessarily have to *originate* from 'altruistic' motives, as long as it develops throughout the relationship into something similarly real as my pleasure when playing a game. We can work on forgetting about the unfortunate, enforced beginning of our relationship and continue on the basis of the (originally unintended) but honest feelings for each other." But imagine finding out that my wife is actually not a human being at all, but an extremely advanced robot that has been sent from the Martians to get information about the human species. She again tries to convince me that she is really in love with me, and that she has developed honest feelings. The disruption in this case seems to be much more *profound*, and the return to a normal relationship is hard to imagine: her assurance that she is really in love loses its force and persuasiveness even if I had certainty (which we can never have about other minds) about the existence of her subjective experiences and emotions. Everything she ever did seems to be worthless, because even if I have benefitted from her behavior by gaining satisfaction and fulfilment, she has not done it for me.[75] She has not done anything at all; it plainly happened because the Martians built her this way and if I had more detailed information about her setup and functioning, I could have predicted that she assures me with exactly the words "she" used. This insight will likely erase all feelings of gratitude (Bieri 2013, 215). Strawson might argue that in particular cases such as this or when we find out that someone has been compelled to an action like the spy, we can temporarily adopt an objective viewpoint and reassess our emotions and reactions. But determinism, he could continue, implies something more fundamental; we had to imagine that *everyone*, including ourselves, is such an advanced robot and, thus, Strawson could argue, we cannot detach us from ourselves and *all* our commitments to others. He could suggest that we will start debating, for instance, about how to deal with the situation, the new insight into determinism's truth, and that by doing so we take each other as disputants with certain rights and duties as serious as ever before. Such would be a perfectly "natural" reaction in Strawson's terms. However, as mentioned before, it is also conceivable that we lose control of ourselves, of each other, and our feelings, and establish a radically different form of living together in a community than currently comprehensible—a community that cannot properly be called human society anymore.[76] The arguments presented in the previous chapter

---

75 Here, I disagree with Derk Pereboom who writes: "Love of another involves fundamentally, wishing for the other's good, taking on her aims and desires, and a desire to be together with her, and none of this is endangered by hard incompatibilism." (Pereboom 2007, 122) This makes little sense. Love is mutually enforced between the lovers through their compassionate behavior, and if my reasoning in chapter four is correct, it makes no sense to say about one's partner that he or she wished for the others' good (and have accompanying feelings of commitment). If determinism were true, then he or she did not do anything; it merely happened.

76 Clearly, it is hard to imagine a human life without reactive attitudes. It would clearly, as John Fischer writes, be "very different from the way we currently understand our lives [...]." (Fischer 2007, 46) We could imagine a scenario that resembles the one outlined in Frank Herbert's Hellstrom's Hive from 1973. The novel tells the story of an underground hive led by a scientist who is convinced of the effectiveness of insect societies. He manages to transform the inhabitants of his hive

underline, as I believe, libertarianism's "seductiveness" in virtue of the facts that we cannot make sense of agency and authorship without genuine free will, and that we are at least as attached to our incompatibilist intuitions than to our moral sentiments (Wallace 1994, 58). Thus, on this basis and in opposite to Strawson, one can have the plausible belief that determinism's truth would equally cause a severe disruption of our moral sentiments. There is evidence on both sides.

Conclusively, I consider this a *stalemate* between incompatibilists and compatibilists regarding the possibility of abandoning the whole set of moral sentiments and reactive attitudes, and because of this, it is reasonable to search for safer grounds in other streams of thought in the debate, namely those that stood in the center of the previous chapter. As long as *these* arguments are not refuted by Strawson, his agnostic compatibilism is in a weaker position than our proposed incompatibilism and discussing these alternatives is, hence, anything but irrelevant.

5.5   Explanation and Intelligibility

Nagel presents another argument in VFN (that cannot be found in ML) that is related to the so called intelligibility problem that was mentioned in the previous chapter. The intelligibility problem is a challenge for an understanding of genuine freedom as unconditional, undetermined freedom. Peter Bieri calls such an idea of unconditional freedom the "nightmare of the decoupled will." (Bieri 2013, 81)[77] If decisions and actions are without preceding causes (such as character traits that determine their occurrence), their occurrence or non-occurrence must be a matter of chance. This does not increase but rather diminishes an agent's freedom. Many of the previously considered compatibilists are eager to harmonize freedom and determinism by emphasizing the ongoing role that sanctions and punishment play

---

physically and mentally in order to create a new species, meaning that each individual is whole-heartedly dedicated to the success and longevity of the hive. There are not special relationships, no inherent motivation of the individuals to do well, but (emotionless) sanction when these beings act ineffectively. Even this scenario might be far from the reality.

77   Own translation

for our system of morality and by other arguments. The reluctance of many compatibilists to accept an idea of genuine freedom stems from the problem of intelligibility, which motivated them to provide other grounds to justify or emphasize the meaningfulness of the aforementioned well-established belief systems and responsibility practices (Hume 1975, 96; Mackie 1990, 226). The previously discussed approach of Peter Strawson is motivated from an opposition against "the obscure and panicky metaphysics of libertarianism." (Strawson 1962, 24) The obscurity Strawson is speaking of concerns the problem of intelligibility: an unconditional free will deprives an agent as much of true authorship as determinism *ex hypothesi* would, so it is assumed. Daniel Dennett calls the idea of indeterminate free will thus a pure "metaphysical curiosity." (Dennett 1990, 138) Similarly, Alfred Mele argues that "indeterminism opens up alternatives [...], but it also raises a very specific worry about luck in connection with moral responsibility. If there are causally undetermined or indeterminate aspects of a process (e.g., a deliberation or an effort to resist temptation) that issues or culminate in a choice—aspects that are present at the very time the choice is made and are directly relevant to the process's outcome—then, to the extent that the agent is not in control of these aspects, luck enters the picture in a significant way." (Mele 1999, 280–281) In a preceding footnote to this passage, Mele mentions Nagel's discussion of this problem. According to Mele, Nagel provides "another useful formulation" of the "luck-objection" in his *View from Nowhere* (Mele 1999, 278). Also Geert Keil names the intelligibility problem and Nagel's formulation in the same breath and discusses them together (Keil 2007, 111). Both the intelligibility problem and Nagel's explanatory problem suggest that we cannot make sense of action without reference to causes external to the agent. I will show in the following that both notions rest on a very narrow understanding of how actions can be rationalized. They presume that only causes can provide adequate explanations of actions, and that it is insufficient to explain actions with reasons. Particularly in Nagel's exposé of the problem, we find formulations that indicate the foisting of causality as a necessary condition into action explanations. Such a move, however, begs the question of free will because it stems from a refusal to accept reason explanations beforehand. Like other explanations, the *adequacy* of an explanation with reasons is dependent on contextual conditions. Causal explanations can be sufficient without denoting the *ultimate* causes of an *explanandum*. Similarly, reason explanations are adequate only when they fulfill conditions that vary depending on, for instance, the (moral) significance of the decision to be made and one's personal relation to it. Similar to scientific explanations or philosophical justifications the search for *ultimate* reasons for actions must come to an end at a certain point. The condition of sufficiency—the question when an adequate point to stop reflecting is reached— is itself a target of rational criticism and evaluation. The dependence of the adequacy of reason explanations on external contexts cannot be set out in detail within the scope of the present enquiry. Instead, I will exemplify the basic idea in the following and show how reference to this framework functions as a proper response to the intelligibility problem.

Let us begin with the classic formulation of the intelligibility problem and thus recall some of the theses defended in the previous chapter. The libertarian account defended takes it to be central for responsibility and freedom that agents have the genuine ability to act. This ability includes doing things, omitting those things or doing other things instead. The ability to act otherwise is already implied in the ordinary notion of agency. However, this invites for the following criticism: If we presume that this means a person could have acted otherwise under exactly the same conditions, this makes her action dissociated from her deliberation and choice. The libertarian seems to conceive such a person as being able to do φ or non-φ despite concluding her deliberation with a preference for φ. In Kane's writings, we find a classic formulation of the problem:

> If my free choice is really undetermined, that means I could have made a different choice given the same past right up to the moment when I did choose. That is what indeterminism and probability mean: *given exactly the same past*, different outcomes ("forking paths") are possible. (Kane 2007, 23)

We can exemplify the problem as follows: Imagine a person P who is on her way to an important meeting and already late. She recognizes an elderly person Q who struggles crossing the street and might get hurt by passing vehicles. P deliberates the significance of her meeting, how much she will be delayed and the amount of danger Q faces (probability and severity of being hit). She also considers how many other people are around but realizes that other pedestrians do not recognize Q's misery. She ends up thinking that it is important to be a good exemplar, and that her chief will accept her being late for helping other people. If this is her state of mind after deliberation, and we apply the "acting differently given exactly the same circumstances"-condition, we must suggest that P could *still* either help Q or not. It seems that she can still either help her or not after having acknowledged that it is prefereable to help Q, if her actions are undetermined. Such an idea makes P's action—whichever it will be in the end—look like the result of chance (Bieri 2013, 286). Clearly, a libertarian cannot be interested in such form of "unconditional freedom."

Thus, a straight forward libertarian response is a *reinterpretation* of the situation and one of the first features to be scrutinized is the *moment* for applying the "could have acted different"-condition (Keil 2007, 114). The argument as presented presumes that libertarians are forced to believe that whatever P's is doing can be detached from P's preference for φ. It locates the opening between the preference or wish, and the actual action. Even stronger conceived and similarly unintelligible is locating the gap between the deliberation process and its results, the formation of a wish or preference. It is suggested that libertarians must assume that if an agent has undergone a deliberation process the outcome of her deliberation is entirely open. Kane exemplifies this as follows: "Imagine, for example, that John had been deliberating about where to spend his vacation, in Hawaii or Colorado, and after much thought and deliberation, had decided he preferred Hawaii and chose it. If the choice was undetermined, then exactly the same deliberation, the same thought process, the same beliefs, desires, and other motives – not a sliver of difference – that led up to John's favoring and choosing Hawaii over Colorado, might by chance have issued in his choosing Colorado instead." (Kane 2007, 23) This would clearly be a strange thing. A person who acknowledges the persuasiveness of arguments in favor of a certain choice and chooses a contrary one instead, cannot be considered rational. Likewise, a person who acknowledges the persuasiveness of arguments for a certain action, decides to perform it, and *does* something contrary cannot be considered rational either. We would clearly doubt that she has really acknowledged the strength of the arguments, or that she really wanted to do the action in question. As mentioned in the previous chapter, it is an integral part of being a rational person to be generally consistent regarding one's preferences and actions (I say in general, because we are familiar with the phenomenon of weakness of the will which is a topic too wide to be discussed here

(Davidson 1980c; Bieri 2013, 100)). Due to this general consistency, it becomes largely predictable what people do. People who proclaim Hawaii to be the greatest holiday destination are inconsistent when going to Colorado instead. The libertarian response to this issue is a refinement of the "acting otherwise"-condition. Libertarians do not care about deliberations, which can result in *any* choice that was considered by the agent. Neither can a libertarian be interested in how *any* action can follow on whatever choice has been made before being it in favor of the action or not. Agents that can act otherwise means that they can continue deliberating and arrive at other conclusions than those mentioned before: the "acting otherwise"-condition must be understood as the ability to *continue* rethinking the arguments for and against a choice and the ability to recognize the greater persuasiveness of previously neglected reasons (Keil 2007, 114). Genuine freedom becomes intelligible when we refine our understanding of when the "being able to act differently"-condition applies. The point of application cannot be after the agent's wholehearted embrace of a choice and after regarding all available alternatives as unacceptable. However, *until* this point the agent can compare the alternatives and the reasons against and in favor *until* determining what to do.

Take again P's decision to help Q to cross the street, because she thinks it important to be a good exemplar and assumes her chief will accept her delay for such a reason: Before she made this decision, P could have held on for another moment (remember that I interpreted the ability to suspend as an action which agents are capable of doing (Locke 2008, 142)) and reconsidered the pros and cons of the alternatives. This is *reasonable* because there are arguments in favor of continuing to go to work, and she was *aware* of those opposing reasons. Until she made her final choice there are *meta-reasons* of the previously mentioned sort that can persuade P to continue deliberating. We are extremely familiar with such process. Uncertainty about the weight of opposing reasons and an insight into the significance of the decision to be made, gives us *another* ground to *continue* deliberating. This is an entirely rational and, thus, intelligible action. Nagel is not satisfied with the description delivered so far. He articulates the following concern:

> If autonomy requires that the central element of choice be explained in a way that does not take us outside the point of view of the agent (leaving aside the explanation of what faces him with the choice), then intentional explanations must simply come to an end when all available reasons have been given, and nothing else can

take over where they leave off. But this seems to mean that an autonomous intentional explanation cannot explain precisely what it is supposed to explain, namely *why I did what I did rather than the alternative that was causally open to me.* It says I did it for certain reasons, but does not explain why I didn't decide not to do it for other reasons. It may render the action subjectively intelligible, but it does not explain why this rather than another equally possible and comparably intelligible action was done. [...] At some point this question will either have no answer or it will have an answer that takes us outside of the domain of subjective normative reasons and into the domain of formative causes of my character or personality. (Nagel 1986, 116)

Nagel emphasizes that our rendering of P's action of both forbearing her decision and continuing to reconsider the other alternatives as reasonable. This argument, however, faces the following objection: We cannot explain *why* P is convinced that this decision is significant enough to continue deliberating. This, we assumed, has been her meta-reason, and there are reasons of which she might have been aware that speak against the significance of the decision and, thus, against the necessity to continue deliberating. Why did P not find those reasons more compelling? Did we really advance the libertarian notion of free agency or just reposition the problem? As a result of his analysis, Nagel suggests that we are taken "outside of the domain of subjective normative reasons and into the domain of formative causes of my character or personality." (ibid., p. 116) This is a familiar step in Nagel's reasoning. Thus, we might start rationalizing her acknowledgement of the significance of the decision by mentioning some psychological traits that make this acknowledgement intelligible such as her general *indecisiveness*. We might say that P had already a hard time choosing her job and holiday destinations and the present issue is clearly just another instance of her disposition. Thus, we offer an explanation that is also suitable for the compatibilist. The character traits determine what P does and make her behavior largely predictable. As argued in the previous chapter, such psychological explanation does not capture this decision in its nature as an *action*. Furthermore, it was argued that character traits are under one's control as outlined in the section on constitutive moral luck: no "normal" agent is *determined* to be indecisive. How then can the libertarian respond to Nagel?

Two objections that are closely related can be put forward. When Nagel writes that we are taken outside of the domain of "subjective normative reasons," he *presupposes* that any intelligible explanation of P's decision must refer to causes external to her. This is not a proof of the unintelligibility of genuine freedom, but rather shows that Nagel is unwilling to accept reason-explanations as sufficient explanations. In contrast, I argued that reason explanations do make actions intelligible, not only subjectively but also from an external point of view. Following up on this suggestion, we can turn around the burden of proof and ask: Why do we demand a *further* reason for P's decision? Choosing *means* deciding to prefer some reasons *instead of others* and this process is intelligible to us as bystanders and observers *when* the reasons are *good*. Clearly, as Nagel writes, "subjectively" many reasons can be motivating. In order to motivate, the agent has to consider them primarily as being persuasive and good: the reasons must compel and move *her* to do this or that (Bieri 2013, 284). If this were not the case we find again a detachment of the person from her actions or wishes as in the case of the alcoholic, who wills something for good reasons but cannot resist her urge to drink (Bieri 2013, 102).

The issue Nagel raises concerns from the observer's perspective on P's action, and I expressed before that from such perspective an action is intelligible if the reasons are good. Consider, for instance, P's situation again and let us refine this example slightly to qualify this idea. Applied to the original case we might—in the spirit of Nagel's argument—be puzzled about *why* she thinks the decision is significant enough to continue to deliberate, and thus, search for further reasons or external factors that affected her decision. We remember that we imagined how she puts "significance" forward as *her* "subjective" reason. We might be aware of strong reasons against her judgment and wonder why she has not seen that the decision is actually way less significant than she believes. Thus, her choice seems unintelligible. But imagine that she could not only help Q to cross the street but save a thousand lives by pressing the button of the traffic light. If she decides to press the button instead of joining the meeting in time, there should be no question about whether that has been a meaningful or intelligible choice. If her decision is questioned, P will argue that the lives of a thousand obviously weighs much more than her meeting and her career, and this is so compelling that no further questions should arise as to what moved her to press the traffic light button. This reason is so *persuasive* that it makes her decision *intelligible*. Thus, we see how intelligibility rests on the quality of reasons. Asking for why she has chosen this alternative instead of another one is absurd. Sometimes we cannot make sense of an action because the reasons put forward by a person are either weak, contradict (as in the case of the alcoholic) what she eventually does, or they are utterly misleading as, for instance, when someone believes to have a reason to lick the stones of St. Paul's cathedral in order to taste their color.[78] There are several ways of responding to these different challenges. In the former case we respond with (moral) criticism: We argue that she *should* have acknowledged the weakness of her reasons, and we make her responsible for a *misjudgment* (Keil 2007, 141). In the latter cases, we face irrational behavior, which motivates the search for external causes. We find better explanations for irrational behaviour in psychological theories about mental disorders and addiction. There is a wide range of reasons and actions that are entirely irrational and in need of external explanation. In contrast, some actions are only eccentric and others are *prima facie* intelligible. This invites to a variety of nuanced judgments and responses in dealing with them and the agents that employ them. We are not forced to see them disjunctively *either* as the result of external conditions *or* as the result of chance. We can try to make them intelligible by understanding what a particular type of person knew, what she could have known (which is a judgment informed by examples of people that have similar backgrounds and act differently in analogous situations), and what might have (externally) constrained her better judgment and behaviour.

It has been sufficiently exemplified that Nagel's formulation exaggerates the problem of intelligibility, presuming that there can *never* be a reason that makes an action intelligible. The adequacy of a justification is dependent on the context and this is analogous to the realm of science. While it is ordinarily sufficient to explain a broken window with the hint that the neighbor's boys played a ball game in his yard, such *explanations* cannot suffice in science where precise concepts, repeated observations and experiments sort out the factors that have a regular influence on the conjunction of causes and effects and provide the basis for establishing theories that explain a set of such phenomena. Even then explanations end at a point in which they seem adequate for the interests at hand (Hartmann and Janich 1996, 45–46). A physician explains the falling of a stone with reference to

---

78 A similar example is mentioned by Ulla Wessels (Wessels 2011, 106).

gravitational pull, but does not explain the existence and mechanism of gravity itself, and this is sufficient when the rudiments of mechanics are discussed. The adequacy of an explanation is here as much dependent on external factors as the adequacy of justifications in the context of action, and there are some reasons that are obviously intelligible.

Clearly, not all decision-making situations are as easily dissolved and intelligible as in the cases when someone is able to rescue thousands of people without having to neglect anything close to equal weight. Sometimes people are confronted with decisions where the pros and cons are equally compelling. Moral dilemmas or existential decisions in which someone is, for instance, forced to either give up his life and family or join the resistance against a suppressive regime that maltreats its citizens and attacks other nations, are well-known from history and literature (Bieri 2013, 73–78). What to do in such situations? We stand like Buridan's donkey in front of two equally mellow heaps of hey and starve because we are incapable of deciding which one to eat first (Rescher 1960). It is a common phenomenon that people adjourn making decisions of such magnitude until the accidental change of external conditions (the fall of a regime, for instance) spares them a decision. When people do decide in such situations, such as to join the resistance movement, they are sometimes unable justify why this and not the other set of reasons tipped the balance. Here, Nagel could rightly argue that neither the agent nor the observers can make intelligible why that person did something and not the other thing instead: There were (equally) good reasons for both alternatives. Clearly, in such situations the unexpected occurrence of shame for not fighting like one's comrades can tip the balance for an agent without providing him with a reason (Bieri 2013, 74). Then, even our own motives can be opaque to us. However, doing something that would fulfill Nagel's demand for intelligibility in such situations presumes that there must be an overwhelming reason speaking for one of those alternatives and finding this is *not* a problem of libertarian freedom but for ethics and decision theory (Wolf 1990, 55; Keil 2007, 117). To summarize, it is first important to understand at which moment a person is still able to act otherwise from a libertarian point of view. This is clearly not the case after a decision has been made and wholeheartedly embraced. Second, actions are intelligible if the reasons for them meet a certain quality. What is comprehensible for the agent and motivates her must not necessarily meet objective criteria of rationality. We react to this gap with criticism and other moral reactions, and only if the gap gets too big, we are pushed to search for parameters external to the agent in the realm of psychological or neurological theories in order to explain her behavior.

5.6 Lucky and Unlucky Innovators

Let me summarize the most important points of the present chapter and integrate them into the problem of innovation. One should not understand the problem of moral luck as philosophical sophistry, or man-made puzzle (and even then, it would probably be worth of being discussed). Instead, what Nagel (and Williams) emphasize in their texts is the close link between the appropriate reactions to misbehavior or exemplary behavior and the topic of determinism that concerned us in the previous chapter. By indicating the many *contingencies* that affect actions first by equipping agents with more or less robust character traits and pitching them into certain social and economic environments, which they have not chosen and can hardly alter, and then later by depriving them of the ability to control parts of the world that affect the consequences of their actions, the problem of moral luck seems to undermine agency from a different angle as the theory of determinism does. We have scrutinized this with hindsight to innovation in chapter two: The external conditions, the contexts of innovation journeys give rise to a lack of control and predictability through complexity (Rip 2012). Mike Martin puts the situation in a nutshell: "[...] serendipity is ubiquitous in science." (Martin 2007, 52) It is, therefore, no wonder that the problem of moral luck first discussed in moral philosophy in the 1970s and 1980s found its way into contemporary RRI discourse (Grinbaum and Groves 2013, 137–139; Stilgoe et al. 2013, 1569).

It was important to understand the problem of moral luck in this chapter in its full complexity and first distinguish between consequentialist and constitutional moral luck. Consequential moral luck emphasizes the *discrepancy* between the reactions to and the evaluation of an "unlucky driver" in contrast to a "lucky driver." As a first response, it was argued that this discrepancy between these reactions could be lowered by reevaluating the agent's actual misbehavior with some distance. After such reevaluation, one can acknowledge that it is grave to be reckless in traffic and accept the risk of producing accidents lightly. On the other hand, there is a sense of tragedy when a person becomes the delinquent of a killing due to circumstances beyond her control. Furthermore, there is an implicit agreement about a distinction introduced in the previous chapter between *believing* in someone's responsibility and *holding* her responsible. Restoring justice and remedying damage is certainly among the functions of legal penalties and this builds the fundament to *treat* unlucky and lucky drivers differently at the courts. This distinction also enters note-less into our initial moral *judgment*. The function of reactions— in this case legal sanction and penalty—to restore justice *affects* our beliefs and this is one of the sources of the moral luck puzzle. This reinterpretation of the moral luck problem can be applied to the context of innovation. Grinbaum and Groves conclude their thoughts on moral luck in the context of innovation with a pessimist statement that seem to doubt any kind of evaluation of an innovator's behavior:

> Even as one strives to possess the requisite virtues of the responsible innovator: to bind one's desire, to check ambition by humility, and to maintain both internal interrogation and external dialogue about the meaning of one's actions, there is no guarantee that moral luck in the uncertain future will not mean that one's efforts to act responsibly will not turn out to have unintended consequences. Whatever choices are made, the final verdict on a distinction between responsible and irresponsible innovation is not in our capacity to make. (Grinbaum and Groves 2013, 139)

Groves and Grinbaum are certainly right in emphasizing that there is luck and contingencies involved in innovation processes that make such processes hardly manageable and predictable. This is entirely in line with my own arguments. The question is: How do we take this into account when evaluating innovators? Grinbaum's and Groves' answer is highly pessimistic when writing that it was not in our capacity to tell responsible and irresponsible innovations apart. If we apply my reinterpretation of the moral luck problem, we first see that we can distance ourselves from the *results* of an innovation journey, and first evaluate quality of *choices* an innovator made and the *reasons* she had to pursue a certain novelty. This, as outlined in chapter three, has to take into account the foreknowledge that was available to an agent and the quality and number of reasonable alternatives that were open to him. Independent of the success of an innovative endeavor, we can evaluate the way in which innovators handle complexity, how they approach

challenges, whether the solutions they offer are creative or whether they are motivated by rather selfish or moral ends. This evaluation can initially be detached
from the *reactions* to the innovators behavior, and can furthermore disregard the
result of the innovation journey. Grinbaum and Groves use the intricate example
of Mary Shelley's *Frankenstein* in which the tragic and unpredictable dynamics
from innovation to disaster is told to exemplify the loss of control as the central
theme in the moral luck debate (Shelley 1989, 138; Grinbaum and Groves 2013).
Frankenstein is a perfect example of an "unlucky innovator." He is so determined
and focused in solving the "mystery of creation" that he loses his clear-sightedness
and discernment over which he stumbles upon when finally being successful
(Shelley 1989, 57). This is the moment when Frankenstein starts to realize that his
fantasy never took him further than through the process of creating. Thus, he is
clueless about how to deal with his creature and more shocked than prepared. What
follows after is the interplay of contingent events with tragic ending. There is a lot
so say about the virtues and vices displayed by Frankenstein, (and the people that
have first contact to his creation) which will be outlined in more detail in chapter
seven. Most importantly, instead of reducing Frankenstein to the consequences of
the complex and unpredictable course of events succeeding his innovation, we
propose an evaluation of his *way of approaching* and *dealing* with complexity and
novelty. On this level, we find differences between lucky and unlucky innovators
that matter morally. It becomes clear from this perspective that a reckless, vicious
innovator is blameworthy—lucky or not. Furthermore, we can still consider the
process from the complementary direction of consequences with regard to the
functions that legal and moral *reactions* ought to fulfill (Rachels 1993, 132). When
innovations have positive results on society, such as the discovery of penicillin,
the invention of the steam engine, or the calculator, etc., we can take these effects
as tipping points in order to decide who is being publicly rewarded with prizes
(such as the Nobel Prize) to *encourage* people who work in other creative
branches. Innovators who bring about such terrific novelties might exhibit from
the perspective of the previous argument the same virtuous traits as unlucky innovators, such as Frankenstein, which includes open-mindedness to novelty, creativity, eagerness, and drive in pursuing their goals. However, they were lucky enough
to innovate at the right time in which their community or the general society was
ready for change and pulled the right switches to pave the way for a diffusion of

the innovation, which then affected society to the better. Thus, the consequences tip the balance when sorting out suitable individuals in order to first push them as exemplars and *incentivize* potential imitators, and second to show some gratitude and *restore* the created value by giving something in return. These underlying *functions* of moral and legal reactions sometimes confuse our judgments (Rachels 1993, 132; Smith 2007). Frankenstein is an unlucky innovator who becomes the victim of his own creation. In a certain sense, he is already punished with a bad conscience when realizing that his "mad enthusiasm" has led to the deaths of his friend Clerval and fiancé Elizabeth (Shelley 1989, 232). This conscience shows Frankensteins regret, willingness to change, and an insight into his wrongdoing which affects our way of treating him (Smith 2007, 482) His bad conscience is evidence that he is already in a process of self-reflection and reform. Robert Walton, the seaman who finds Victor Frankenstein in a desolate state in the arctic ice, pities him for "the greatness of his fall" and Walton's only urge is to soothe him from his misery (ibid., p. 266). He sees no necessity for further reprehension since Frankenstein has already been punished enough. Clearly Frankenstein's final words are an expression of repentance, which is nothing else than the recognition of one's individual responsibility (Strawson 1962; Bieri 2013, 363). Nevertheless, Walton could agree with the same critical judgment that Frankenstein expresses when speaking about the "madness" that drove him into blind dedication: Frankenstein is *blameworthy* but the ultimate *result* of the story is not completely his responsibility because these events were beyond his control. Bringing about novelties is not a solitary exercise: Many other hands are involved and we will survey in the next chapter whether and how this affects individual responsibility.

6. Collective and Corporate Responsibility

6.1 The Problem of Agency Revisited

Visioneering was introduced in chapter three as a case study about a relevant agent in the innovation process that increasingly attracts the attention of TA and STS scholars. In contrast to other more "traditional" agents, such as engineers, managers, and politicians, visioneers are in many respects special—as argued before. Let me briefly recapitulate the theses previously presented: Large-scale research like the NNI require the gathering of a variety of stakeholders. Scientific endeavors that require large budgets cannot be instantiated by single individuals. They require coordinative and communicative boundary work, for which conceptual commonplaces that articulate shared values and goals and, in some instances convince the public of those goods, are helpful: Visions fulfill those functions and are utilized for such purpose. We have learned from the history of technology that technologies are socially constructed, and that we can establish complex narratives of their development (Bijker et al. 1987; Nye 2006; Rip 2012). We discovered that visioneers operate in a variety of different *contexts* and *social environments*. Amongst those are more loosely structured institutions or public spaces and more "regulated" and organized environments, such as companies, and universities. This gave rise to the question of whether institutional impact and the divergences of those contexts affect individual responsibility (see chapter three).

Can we distribute responsibilities in highly complex arrangements in which many agents contribute to an overall phenomenon, event, or attribute a general responsibility to such entities as a whole (Lenk 2007, 199)? It has been described in chapter three that visioneers are an important factor in this process. Their involvement in the technological development takes place on various levels and through a variety of activities. As McCray mentions, visioneering activities entail the promotion of visions through authoring books, presenting them and the visions outlined in them at conferences and other public forums, individual networking, entrepreneurship, scientific, and engineering activities (McCray 2013). Some of

© Springer Fachmedien Wiesbaden GmbH, part of Springer Nature 2018
M. Sand, *Futures, Visions, and Responsibility*, Technikzukünfte, Wissenschaft und Gesellschaft / Futures of Technology, Science and Society,
https://doi.org/10.1007/978-3-658-22684-8_6

these activities are rather embedded in strong institutionalized and corporative settings, such as the scientific and engineering work. Other visioneering activities are performed privately, like writing (for the public). Furthermore, the role in which the institutionalized actions are performed might vary. Visioneers can work as typical researchers at a university and, at the same time, as a chief of a company. Consider the example of Ray Kurzweil who was the Chief Engineer of Google, the founder of the Singularity University, an entrepreneur, and a freelance author.[79] Many of these roles have been closely intertwined. On those corporate affiliations of visioneers, McCray notes:

> The visioneers' hybrid nature—a combination of futurist, researcher, and promoter—and the influence they sometimes attain compels us to consider how they interact with other actors in broader systems of technological innovation. Business executives and academics have often employed ecological metaphors to describe places where technological innovation occurs. These complex and dynamic "ecosystems" are home to some familiar "species." These include established companies, universities, law firms, patent lawyers, entrepreneurs, investors, government funding agencies, the media, and, of course, scientists and engineers. (McCray 2013, 14)

Prima facie, the divergent roles that visioneers occupy likely affect their individual responsibility. The first reason that supports this view is how the institutional impact might restrain the alternatives that are open to them. Institutions and corporations might permit or require certain forms of public representation. Visioneers might be required to concur with their universities' or companies' corporate identity when hired. They are probably not allowed, for instance, to join other affiliations or seek memberships in entities that diverge fundamentally from their core values. Visioneers might be required to act in the best interests of their corporations. In some cases this might imply, as indicated in the third chapter, they promote a vision although the content of such vision does not reflect their personal beliefs. Certain leading positions require representatives to publicize the visions

---

79   Some of Kurzweil's achievements are listed in his The Age of Spiritual Machines (Kurzweil 1999, 174–178). The list displays an outstanding legacy that was rightly honoured with the National Medal of Technology by President Clinton in 1999 (McCray 2012, 367).

of the technologies produced by the company and thereby raising public awareness and marketing them. This is merely a professional obligation, and there are few reasonable alternatives to fulfilling this requirement for anyone aspiring to become leaders or companies' spokespersons. Either you identify with a company's corporate philosophy, or you will never reach such a position. If this is your goal, you are forced to deliberate this trade-off.

As also indicated in chapter three: In many businesses the promotion of visions is not only part of marketing the corporate products but also of the usual process of partnering up with other institutions and acquiring project funding by suggesting ideas and future opportunities. This is very similar of creating a shared, inter-institutional vision. In many cases, technological visions are reasonable tools to ebb the way for corporate purposes, setting up innovation pathways and so on. Furthermore, the corporations' climate either encourages or discourages certain behavioral patterns.

In a famous case from the 1970s—the Ford Pinto case—it has been said that Ford's management, represented by President Lee Iacocca, fostered a mentality that is properly condensed in the President's own words as: "Safety doesn't sell." (Lenk 1993a, 198–199) The Pinto was a compact car that Ford delivered, despite awareness of a problematic gas tank. The placement of the tank resulted in an increased likelihood of inflammation when rear-end collisions occurred. A cost-benefit analysis (based on false assumptions) undertaken by Ford revealed that adding a plastic buffer for $11 USD would be more expensive than compensating the relatives of an alleged annual number of 180 traffic deaths plus an arbitrary number of burn victims. Twenty million Pintos were delivered in the following years (ibid., p. 99). The number of people that were killed because of the identified flaw of the gas tank is and remains eventually unknown. Until 1978, the company recalled 1.5 Million cars and stopped producing it altogether in 1981. The public contempt of Ford's "morbid calculous" grew too big. The *Time* magazine ranks the Pinto amongst the top 10 most infamous product recalls and writes: "The Ford Pinto was a famously bad automobile, but worse still might be Ford's handling of the safety concerns [...]." (Various 2009) This case is interesting and, therefore, often discussed in business ethics because it is *prima facie* an example of how moral values or obligations and economic interests can collide. In stark contrast to this analysis, liberals might claim that it has been the public display of the "Safety doesn't sell"-

policy that eventually led to the failure of the car in 1981. Hence, caring for moral values such as safety (at least ostensibly) is, indeed, congruent with one's economic interests. The possible self-interest/economic interest versus moral interest collision is nevertheless one significant dimension of the Ford Pinto case.

There is, however, another dimension to this example that is more closely connected to our topic of collective and corporate responsibility. Ford's leading engineers, who knew about the technical risks of the gas tank, argued that they were not in a position to articulate concerns or debate safety issues openly. They said that such attempts would have resulted in suspension (Rau 1978; Lenk 1993a). The corporate's climate apparently suppressed safety discussions and induced fear amongst staff members. By highlighting this aspect of the corporate policy, the engineers clearly tried to repudiate their individual responsibility. The corporate's climate as a contextual aspect of decision-making is, however, only one element for properly evaluating the engineers' behavior. Generally, as mentioned before, the corporation's statutes and internal climate provide a framework in which the individual agent has to deliberate alternatives, and it is not obvious that fear of suspension is an acceptable reason to leave vast risk issues unexpressed (DeGeorge 1991, 163). In general, engineering is a profession that usually offers a huge number of job opportunities for practitioners, especially in North America and Europe. In some cases, given the possibly devastating and far-ranging results of innovations, it can indeed be reasonable to raise concerns openly and, thereby, take the risk of dismissal. Some of the engineers in question would not have suffered from a loss that is even close to being equivalent to the damage that resulted from the technology in question.

My proposal to this point is admittedly very unspecific for the following reasons. First, this is due to my lack of familiarity with the details of the case—including knowledge about the backgrounds of the persons involved in the Pinto case and the exact circumstances in the firm—to provide a final judgment. It is beyond the scope of this enquiry to shed light on those details. Since the reasonableness of choices varies starkly with the background of the individuals involved, this is a significant deficit for a proper evaluation. As argued previously, many engineers have a vast amount of job opportunities in North America and Europe that makes risking one's job position in the light of certain amounts of damage a reasonable option. However, there might be other people who have found the only suitable position in the area, and who are bound to their families or have other ties that make them dependent and immobile. Then, losing one's job is accompanied with a much higher individual risk. Second, even if we had extended knowledge about the case, it would not be simple to determine individual responsibility. The very superficial case description given above suggests that there were only two alternatives available to those engineers: risking one's job by raising the issue of safety or keeping the issues for oneself and not discussing them at all. This is a *simplification* of the complexity of ordinary decision-making situations. It is un-likely that the company would have suspended all engineers together. Thus, maybe it would have made sense to team up and advance the issue of safety through com-mon effort and, for instance, increase pressure on the board by holding employee strikes.

If such an opportunity has been available, it was clearly false to claim that one has done the right thing by not risking one's job because it was not the only thing one could have done. Furthermore, a safety policy such as Ford's is unlikely to evolve within a short time span. Throughout the years (or even decades) before the Pinto was released, Ford's management had established such practices slowly but steadily. Board members must have backed Iacocca's policy long before the Pinto incident. Note that Iacocca was named Vice-President of the Ford Motor Company and general manager of the Ford division already in 1960. The Pinto case might have just been the culmination of a series of managerial failures. Throughout this period of what could be called the companies' "safety decline," engineers (especially seniors) might have had a variety of op-portunities to intervene and propose a shift in direction. Again, the corporate stat-utes and internal hierarchies possibly hamper the effectiveness of such proposals. Regarding those, however, the same basic idea applies. If there are opportunities to alter internal hierarchies and the general climate and, thereby, make a significant moral change, it would be praiseworthy to pursuit such an effort and blameworthy to neglect it.

As one can see, there are many "ifs," "might haves," and "maybes" based on my reasoning. This is natural regarding my lack of insight in the details of the case. However, the main purpose of this section has not been to provide a definite judg-ment about who has to be held or not to be held responsible in the Ford Pinto case. The purpose of the brief (speculative) exploration was to remind us that corporate statutes, climates, and hierarchies put tight jackets on the individuals in those companies, but they do not undermine their freedom and responsibility. It makes agents in corporate settings co-responsible if they allow opportunities to alter and enhance morally significant decisions of their corporations slip through their fingers, as long as there are no equally weighty reasons in favor of omitting such steps.[80] I am inclined to regard the situation of agents in corporative and collective settings as largely analogous to their situation in innovation processes outlined in chapter three.

When we look at the involvement of agents in corporate settings, we merely consider the dynamics of the system "corporation," just as if we were considering the system "innovation process." In each of these cases, agents are constrained in their choices and their impact on the "overall agenda" but not deprived of their responsibility. Similar to the context of innovation, personal interests, technical boundaries (when working with machines or technologies), corporates interests and statutes, stakeholder demands, and moral requirements put significant constrains on individuals. Responsibility means deliberating alternatives, advancing new possibilities, and maneuvering reasonably through such external constrains. Admirable character traits that contribute to such practical faculty of judgment are, amongst others, creativity, eagerness, and clear-sightedness. I will explore those virtues and their role for handling complexity in more detail in a later chapter.

Although, I have discussed the co-responsibility of engineers, it should be clear by now that the same reasoning applies to the board members, managers and other staff members. Within the range of their alternatives, responsibility requires agents to pursue betterment and to advance the instruments to reach such betterment always in consideration of the possible downsides of such change. If opportunities for betterment are missed out without good reasons, co-responsibility for the resulting failures of the company evolves. Relative to the position that is occupied and the accompanied impact on corporate agendas, this co-responsibility

---

80   The condition "morally significant" is significant. One must not assume that there
      is a strong obligation to change the color of the corporate logo just because one does
      not like green—the color of the logo. When human lives are involved, however—
      as in the Pinto case—the stakes are higher and the risks to be taken to prevent harm
      can have a proportional weight.

can be profound. Co-responsibility means the responsibility for the right and wrong actions of a person in the contexts of collective activities. The collective activities together can produce harm that the individual wrongdoing would not have produced. However, responsibility is carried only for the individual wrong-doing and this *includes* whether and how one takes the possibility of the wrong-doing of others into account. I will come back to this when discussing the Discursive Dilemma in a later section. The term co-responsibility captures these features most adequately. Applying these very general and abstract remarks to real corporate policy making can certainly be complicated and frustrating. It is, unfortunately, beyond the scope of the present book to discuss measures that stimulate and enhance the moral climate in firms, which allows for a more creative development and maintains the company's economic success at the same time.[81] I have presented some suggestions on how to interpret individual responsibility in collective and corporative settings.

There is a second reason to consider collective and corporative responsibility in more detail. In regards to more severe cases, the feeling might grow that individual's effectiveness and, hence, responsibility vanishes altogether. Regarding these cases, it might be more reasonable to attribute moral responsibility to the collective entity as such. Take the following example introduced at the beginning of Philip Pettit's article *Responsibility Incorporated*, in which the problems in determining individual responsibility and the idea of ascribing corporate responsibility are mentioned:

> The *Herald of Free Enterprise*, a ferry operating in the English Channel, sank on March 6, 1987, drowning nearly two hundred people. The official inquiry found that the company running the ferry was extremely sloppy, with poor routines of checking and management. "From top to bottom the body corporate was infected with the disease of sloppiness." But the courts did not penalize anyone in what might seem to be an appropriate measure, failing to identify individuals in the company or on the ship itself who were seriously enough at fault. As one com-

---

81   There is a vast amount of literature on this topic. The author finds (Collins and Porras 2002) recommendable. See also (Martins and Terblanche 2003). I will return to this question and make some indications in chapter seven.

> mentator put it, "The primary requirement of finding an individual who was lia-
> ble... stood in the way of attaching any significance to the organizational sloppi-
> ness that had been found by the official inquiry." In a case like this it can make
> good sense to hold that while the individuals involved may not bear a high degree
> of personal responsibility, together as a corporate enterprise they should carry full
> responsibility for what occurred. (Pettit 2007, 171)

Pettit claims—and defends later on—that it makes sense to hold the staff members together as the corporate enterprise morally responsibility. Can corporations qualify as agents and become an addressee of moral responsibility ascriptions? In my text, I have been concerned thus far with individual responsibility within collective and corporative settings and argued that despite a lack of causal efficiency for the overall agenda, co-responsibilities for members of corporations within reasonable limits can emerge. These responsibilities are limited to the reasonable alternatives of individual staff members to create change. Such responsibility depends on the stage of the hierarchical ladder that someone is located at, since this positioning affects their impact on the corporate behavior. In chapter three and four, I discussed the more general conditions for individual responsibility amongst which were the capacity to balance reasons and act according to those reasons. Series of events that involve the body of a human person, but happen in the absence of fore-knowledge or control—such as stumbling or coughing—are not properly qualified as actions. They must be considered as mere behavior, and their possibly negative results are accidents. In contrast, the corporate entity does not have its own intentions or the capacity to balance reasons, at least at first sight. It merely aggregates a number of varying agents with probably diverging intentions. In short, the corporate body lacks the conditions of being an addressee of moral praise and blame. However, I just referred to corporate behavior, and there is a plenitude of ways in which we ordinarily apply intentional vocabulary to corporate behavior. This fact might pave the way for more advanced arguments that shows that corporations and collectives do qualify for responsibility attribution. Consider this point outlined by Patricia Werhane and Edward Freeman:

> There are a number of senses in which we ordinarily think of firms as moral
> agents. Like individuals, corporations set goals. These goals are often defined in
> a mission statement, delineated in policies, or operationalized in the corporate
> culture and activities in which the corporation is engaged. In ordinary language

we refer to corporations as actors, and we hold them, like individuals, responsible. For example, we say that Ford failed to act when it did not initially change the design of the Pinto despite its knowledge about the unfortunate placement of its gas tank. We praise 3M for its environmental programmes or DuPont for initiating an alliance of chemical companies to improve environmental and social performance. (Werhane and Freeman 2010, 519)

Given this accurate observation about our application of intentional language in our ordinary way of speaking about corporate behavior, we should consider whether this makes sense, or whether there are reasons to abandon such practice, if corporations do not meet the conditions necessary for moral agency. I will argue in the following section that such an ordinary way of speaking is a pragmatic foreshortening often because of our prevailing lack of knowledge about who was really in charge when corporate misbehavior occurs, which alternatives where open to the involved agents, and what their backgrounds are. In short, the ordinary speaking of corporates intentional behavior is a foreshortening justified by our initial lack of insight into the complexity of the decisional situation. In contrast to this pragmatic theory, Peter French argues that it is reasonable and possible to attribute corporate intentions and, thereby, infer on their corporate responsibility. He sets out an advanced argument for corporate personhood. This approach will be discussed and criticized later on in this chapter, but first I will assess the existentialist account of collective responsibility brought forward by Karl Jaspers and Larry May. I will argue that the existentialist account of collective responsibility is merely an account of individual responsibility in disguise.

6.2 Collective Responsibility—The Existentialist Account

Karl Jaspers most likely discussed the problem of collective responsibility for the first time after World War II. At the beginning of the Nuremberg trials, Jaspers thought and wrote systematically about the collective guilt of the Germans for the horrors of the Holocaust and the starting of the war (Jaspers 2012). His viewpoints, which were presented through lectures to the public in 1945 and published afterwards in the book *The Question of German Guilt*, had a huge impact on the debate about collective responsibility in the following decades (May 1991, 239).[82] Hence, it is reasonable to consider his ideas in this section first briefly. Jaspers distinguished four forms of guilt: political, criminal, moral, and metaphysical guilt. By distinguishing these concepts, Jaspers wanted to contest the many uncritical accusations that were directed towards the Germans as a collective after the war. He argued that the legal punishment only covers political and criminal guilt. Political guilt should be understood as everyone's liability for being a member of a state. If a state gets involved in an international conflict, the winners of the conflict are allowed to put charges upon that state. Every individual member of this state has to bear these charges whether or not all of the individual members consciously or unconsciously contributed to the rise of the conflict. Hence, Jaspers argued on this level that the reparations that the Germans had to provide were justified. But this kind of political liability does not imply that everyone is morally or metaphysically guilty for the faults of this state, following Jaspers.

The second form of guilt, criminal guilt, concerns those individual actions that are against the law. The instances of accusation for criminal guilt are the courts. The legal response to criminal behavior is punishment. Moral guilt, in contrast, concerns particular individual actions. Those actions, even if there is no legal representative questioning them, are tested by one's individual conscience (Jaspers 2012, 19). Repentance and renewal can follow an internal process of reflecting one's behavior and considering one's own conscience (ibid., p. 23). Last but not least, Jaspers argues that metaphysical guilt arises from the solidarity between humans. The instance that judges metaphysical guilt is god. Jaspers own argumentation is sometimes dangerously unclear. At the end of the paragraph in, which he distinguishes the four types of guilt, his framing of the concept of metaphysical guilt shows similarities to the Christian understanding of the original sin, a sin that is inherited by birth. A slightly different argument centered in Jaspers' argument

---

82   Larry May's revitalization of Jaspers' argumentation was adopted and further employed in applied ethics. Hud Hudson, for instance, used it to defend that meat consumers' are morally tainted regarding the cruelties committed to guarantee meat supply. He proposed this as an alternative way of framing collective consumers' responsibility thereby contrasting the influential utilitarian approach to vegetarianism (see (Hudson 1993)).

claims the *inevitability* of situations in which the only way of helping someone would be risking one's own life with little or no probability of success. Jaspers assumes that the situation in Nazi-Germany meant similar situations for many inhabitants who condemned the regime and its proponents. They did not risk their lives, which could have been ineffective. The attribution of a moral guilt in such cases would be inappropriate according to Jaspers.

However, these persons bear a metaphysical guilt "for being still alive." (p. 20) Inside the feeling of shame for the failures of the community, metaphysical guilt finds its expression (p. 21). Larry May, who adopted and extended Jasper's concept of metaphysical guilt, also highlights the feelings we have towards the general achievements and failures of the communities by which we live (May 1991, 243). For him, as an existentialist, it is important to note that we are not only responsible for our actions but also for the decision of what kind of character we become throughout our lives. When someone is covered with the moral taint of his community, it is up to them to change their *attitude* towards this community and condemn their faults. Metaphysical guilt implies that one should dissociate themselves from the "feeling" of being part of such a community and, instead, try to become a person whose identity is no longer associated with it. This process does not necessarily have to have a causal implication. In contrast to the typical understanding of moral guilt, May writes:

> The guilt is metaphysical because it has to do with who one chooses to be, in the sense of whether one accepts or rejects such affiliations, rather than what one has actually done. [...] Moral guilt is appropriate when there is something that a person brought about in the world. Being a causal agent is closely connected to moral guilt. But when a person's causal agency is not in question, or at least when the causal role one played did not make a difference in the world, then moral shame or taint may be the appropriate moral feeling. We are thrust into the domain of responsibility for who one is, and the terms of judgment should be different than those which are addressed to a person's explicit behavior. (May 1991, 247–248)

Both Jaspers and May contributed to the discussion of collective responsibility by pointing out that there are different kinds of responsibilities (or "guilt" in Jaspers' terms) in collective settings. They noticed that a complex collective is not just the sum of the individual actions although they can be constructed as if a mathematical division is possible. Collectives lack all the attributes that have been discussed before, that constitute individual responsibility (freedom to want and do otherwise). It is neither an autonomous subject, nor can it respond and justify its behavior. Meaningful response and justification are always reactions of subjects that are able to take reasons into considerations. In separating (political) punishment and moral guilt, Jaspers does justice to the distinction of the individual actions and the complex interrelation in which it is performed. Their emphasis of the role of attitudes and character traits also resonates very well with my reasoning in previous chapters and with my virtue ethical point of view. It was mentioned, for instance, in chapter three that in order to explain the moral sentiments that forgetful and reckless agents arouse we have to refer to a sort of responsibility for attitudes. In cases of forgetfulness, the behavior of an agent is not causally effective, yet something different and presumable better had happened, had they taken up the task of alternating their attitude in a forward-looking manner. However, besides their large disservice to distinguish different forms of responsibilities, May's and Jaspers' notion of metaphysical guilt, and their argumentation is flawed in several respects. Before I discuss their approach critically, I must point out another central feature of their theory. There is a prescriptive component attached to their notion of metaphysical guilt that is not properly articulated in the passages cited above. May also wants us to distance ourselves from group harm:

> [...] metaphysical guilt arises out of each person's shared identity, out of the fact that people share membership in various groups that shape who these people are, and that each person is at least somewhat implicated in what any member of the group does. But metaphysical guilt is not merely based on group membership. Rather it arises out of the fact that a person did not but could have (and should have) responded differently when faced with the harms committed by his or her fellow group members. The guilt arises from the fact that nothing is done to prevent the harms or at least to indicate that one disapproves of them. Due to these failures, the individual does nothing to disconnect himself or herself from those fellow group members who perpetrate harms. And while it may be that one is also morally guilty for one's omissions which contributed to the harm, what is important for metaphysical guilt is that one chose to do nothing to distance oneself from the harmful acts of one's fellow humans (or more plausibly, some smaller subgroups of humans). Metaphysical guilt is based not on a narrow construal of what one does, but rather on the wider concept of who one chooses to be. (May 1991, 240–241)

We can summarize this position as follows: Given that one is a member of group that is doing harm to others or committing faulty actions, feeling tainted or guilty is appropriate, even though one could not have influenced the overall behavior of that group. I will outline some objections to this conception in the following paragraphs.

First, the type of communities in both Jaspers' and May's analysis are fairly unspecific, which entails some practical and theoretical difficulties. May discusses a university community, Jaspers talks about nations. These two types of entities already differ in important respects (van de Poel 2015b, 56–61). Taking classes at a university or holding a chair as professor presumes the conscious agreement or even the previous application for such a position. Becoming part of a nation, in contrast, is something, which is not your own decision in the first place. You are born into a national community, whether you like it or not (Nagel 1991a). Later on, you might regret that you never became a citizen of a different state, but being born as an American or a German is not something you consciously decide. These differences matter insofar as the identification of an agent with his community can be strengthened or weakened with regard to such a decision. When becoming part of an animal rights association because of your willingness to support the general goals of such an association, one's identification does not rest on a sense of belonging to the community. The identification is presumably already given when signing up. If the association is in an essential respect changing its main goals, then it is up to the members to resign. The equivalent in regard to citizenship is not equally simple. Leaving a country or adopting a new nationality is not as easy as applying and inscribing as a student for university or another sports club. As I indicated at the outset of this chapter, the general point is that one must clearly separate the different settings in which collective actions are undertaken. Concerning moral responsibility, it makes a huge difference whether you are part of a more corporative institution in which dropping out is easier; associations, political parties, or harder; companies with binding contracts. Typical corporate institutions, such as companies and associations, are internally structured in a way that does justice to the degree of influence of the individual members on the overall agenda. In large institutions, it is rarely the case that normal employees have a saying regarding the major strategies and goals. That is up to the chiefs. If such an internal hierarchy exists, it is unreasonable to neglect it concerning the responsibilities of the individuals in such corporate enterprises. Neither has May nor Jaspers taken these differences of institutional acting and limitation into account. Hence, their general approach needs elaboration. Through the analysis of some examples, Hans Lenk and Günther Ropohl have shown how a multilayered concept of responsibility can be applied to the complex arrangements in which engineering actions take place (Lenk 2009; Ropohl 2009). Lenk and Ropohl sketch exemplarily how one could interpret and evaluate alternatives for actions (resistance or dropping out) in institutional settings on different levels of hierarchy. These distinctions apply even more to the case of visioneering. Visioneering—as outlined in chapter three—is located in different institutional settings, and the individuals may well be found in various positions in these settings, for example: as chiefs, normal employees, or principle investigators. A different set of responsibilities apply for each of these positions.

As mentioned earlier, May's and Jaspers' way of framing is the definition of community remains too vague to function as a foundation for their thesis of metaphysical responsibility. If we think of the numerous non-chosen communities we are living in, it becomes apparent that the concept of metaphysical guilt inherits a lack of plausibility. As a German, for instance, I am not only a citizen of a state but also automatically the member of the German-speaking language community that is spread all over the world. Sometimes I am attracted to the many impressive literary products full of rich metaphors and grammatical finesse that have been written by members of this community. Hence, the main condition (being member of a community and sharing an emotional bond) of being morally tainted by the faults of this community are the conditions met. If sharing a language means being part of a community, then I am tainted by the crimes of the German-speaking people all around the world although I share nothing with them other than their language. Now, concerning my second point previously mentioned, it is hard to see which reaction would be appropriate in this situation. Apparently, one cannot get rid of one's mother tongue. If I am obliged to develop a dissociated stance, as Larry May asks us to get rid of the moral taint, the distinction between the meaning of metaphysical guilt and the meaning of being a moral agent in general is blurring to inseparableness. Most people would agree that condemning the evils in the world, at least mentally, is part of being a moral subject that assesses human practices and characters according to values. If this is what Larry May demands us to do according to our membership in communities, then arises the question: Why this should be considered as a special type of *collective* responsibility?

It is crucial now to understand the point, which could disprove Jaspers' and May's argument, although not in a sense that proves their reasoning to be invalid. Rather, it can be shown following up on this that they miss the target of their proof: The existentialist way of reasoning about collective responsibility is threatened by the just outlined problem of vagueness in May's and Jasper's concept of "group membership" and the accompanied ubiquity of such notion. How exactly is this aspect threatening? The fact that one does not choose group membership and is *nolens volens* a member of the group of human beings makes one responsible for the harm humans do. Global harms—according to existentialist reasoning—like massacres in Africa, wars in Sudan and the right infringements in Russia taint *all humans*. Responsible conduct, according to May's reasoning, requires one to react to such group harms by distancing oneself from such harms. Let us recall what May understands as the appropriate reaction regarding collective wrongdoing:

> An existentialist account of responsibility can be rendered initially plausible by noting that people often feel pride as well as guilt for what their fellow community members have done. The sheer fact of one's membership in a biologically or geographically defined group is normally not under one's control. But what is under one's control, and hence the subject of appropriate feelings of responsibility, is **how one positions oneself in terms of that group**. In communities, as in other groups, it is **not** mere membership that creates responsibility, but **how one reacts** (behaviorally or attitudinally) to the groups of which one is a member [own emphasis]. (May 1991, 244–245)

The key to understand existentialist collective responsibility is expressed in this passage. What May provides in the above quote is a reasonable account not for collective responsibility with regard to group membership, but a reasonable account that expresses our expectations towards responsible individuals for developing an attitude regarding group harms in general, independent from membership in a blameworthy group. If May has attempted to show that there is a collective responsibility that is based on the condition of group membership plus appropriate moral attitudes towards one's fellow group members, he has missed his target of proof. What he provides is the reasonable condition that one should disapprove group cruelty, whether that person is a member of that group or not. People who are conscious and know about the Holocaust or the My Lai massacre, if they wish to be credited as a moral character with true moral motivation and intention they should despise such cruelty, publicly or in private for themselves. This idea is extremely appealing; however, it constitutes a theory of *individual* responsibility *regarding* collective cruelty, but not a theory of *collective* responsibility.

6.3   Corporate Responsibility—French's Account

As previously explained, there are vast differences in composure and "nature" between (random, unstructured) collectives and corporations.[83] In contrast to random collectives like demonstrators or the German-speaking language community, there is a number of entities that usually maintain a legal standing. These include corporations, associations, NGOs, (some) civil society organizations (CSOs), companies, universities, churches, etc. I have mentioned before that corporations, just like associations, universities, sports clubs and other legal bodies, have statutes and clear guidelines regarding decision-making, hierarchical structures, verbalized and designed corporate identities, as well as many others. Corporations are established, for instance, to create shareholder value, to produce artifacts or to help satisfying consumer demands in an environmentally friendly way. The variety of such purposes is practically infinite. In his landmark paper *The Corporation as a Moral Person*, Peter French provides an argument for corporate responsibility based on these characteristic of the decision-making procedures established in almost all bigger companies. These structures—according to the argument—allow us to treat corporations as "full-fledged moral persons [which can] have whatever privileges, rights and duties as are, in the normal course of affairs, accorded to moral persons." (French 1979, 207) We should assess this argument in detail. French starts by distinguishing legal, moral, and metaphysical personhood at the beginning of his article (ibid., p. 207). The first crucial step in his argument that sparks our interest is the distinction of two types of responsibility (p. 210). The first type is understood as causal responsibility "[which] is usually used when an event or action is thought

---

[83]   The Mafia is in this respect a difficult case, located somewhere in between.

by the speaker to be untoward." (p. 210) This notion of responsibility merely stops after sorting out from which subjects (or objects) the events that are disapproved stem. The certain objects that can be determined as the originators of undesired events are responsible in this first sense without necessarily being responsible in the second sense. French adopts the second sense of responsibility from Elizabeth Anscombe and understands it as the "liability to answer." This second type of responsibility ascription goes hand in hand with the first:

> A responsibility ascription of the second type amounts to the assertion of a conjunctive proposition, the first conjunct of which identifies the subject's action with or as the cause of an event (usually an untoward one) and the second conjunct asserts that the action in question was intended by the subject or that the event was the direct result of an intentional act of the subject. (French 1979, 211)

If the subject in question has intended to execute an action φ and is asked by someone to answer why she has done so, she can—regarding the alleged purposefulness of her behavior—react with sharing the reasons for her action φ. In many cases, such answerability is the result of a legal relationship: Courts demand answers from defendants in legal cases in order to set out whether there should be a conviction or not, and to determine the degree of penalty. In cases of implicit or explicit contracts between business partners, lovers or friends, such answers can also be demanded. Regarding the general demands of morality, French argues that they "hold reciprocally and without prior agreements among all moral persons. No special arrangements needs to be established between parties for anyone to hold someone morally responsible for his acts […]." (French 1979, 211) We can even extend this proposal and argue that the demands of morality, for instance to be honest, even hold if no know can possibly know of a committed deception or fraud. By regarding morality, so to say, we can employ some sort of fictitious instance to which, in principle, must answerable regarding our reasons for action. So far, French's reasoning largely carries the spirit of my own arguments introduced in the previous chapters, with the exception that the assumed relation of answerability adds another layer to the analysis. But such notion is easily covered within my framework. My proposed condition of "balancing reasons" in chapter four explains such requirement quite well. Answerability is merely the capability of expressing the reasons one had for action. Answerability itself, therefore, rests on the condition of reasonable intentionality as outlined before. It does not constitute another distinct condition that must be presumed, it is merely an extension of the basic condition of deliberate intentionality. Now, French departs from this point in his paper into a different direction. He assumes that since certain subjects having legal rights such as children or animals, which have the right for inviolacy and lack the capacities required for the second type of responsibility ascription, they are no full-fledged members of the moral community. Hence, legal personhood, as French argues, does not imply moral personhood. This also applies to entities that are composed of a number of agents such as corporations. French claims against Locke that legal personhood does not imply moral personhood, which naturally follows from the previous reasoning. Therefore, to be an appropriate addressee of moral responsibility—to be a moral person—presupposes that the subject is an intentional agent. In setting this out, French adopts Donald Davidson's definition of agency. Davidson writes in his article on *Agency*:

> In the case of agency, my proposal might then be put: a person is the agent of an event if and only if there is a description of what he did that makes true a sentence that says he did it intentionally. [...] If we can say, as I am urging, that a person does, as agent, whatever he does intentionally under some description, then, although the criterion of agency is, in the semantic sense, intentional, the expression of agency is purely extensional. (Davidson 1980b, 46–47)

This proposal is based on the classic analysis of Shakespeare's Hamlet, which I also discussed in chapter two to introduce intentionality as a precondition of responsibility ascriptions. Davidson rightly argues that there is a description of Hamlet's behavior that says he did it intentionally. It is wrong to describe the events that involve Polonius death as: "Hamlet killed Polonius intentionally," because this is not what Hamlet did intentionally. However, killing the person behind the curtain, which he did intentionally, turned out to be Polonius. Davidson's account of agency captures this ambiguity by proposing two conditions for agency: first, that there is a description of the events that says that X did φ intentionally, and second, that "X did φ intentionally" is a true description of the events. Hence, the description "Hamlet killed the person behind the curtain intentionally" satisfies both conditions. This allows for the conclusion that Hamlet has been the agent of the act of killing the person behind the arras, which makes him responsible for this. French transfers this account of agency into his theory of corporate agency:

> For a corporation to be treated as a Davidsonian agent it must be the case that some things that happen, some events are describable in a way that makes certain sentences true, sentences that says that some of the things a corporation does were intended by the corporation itself. (French 1979, 211)

This is French's proposal for corporate agency. My first remark regarding this reasoning concerns the relation between agency and moral personhood. Whether Davidson's notion of agency really captures the notion of moral *personhood* could be questioned for good reasons. There might be people who occasionally satisfy Davidson's conditions of agency without being moral people. This is caused by personhood and might also require coherent behavioral patterns over time, some lasting ground projects, and an (underdetermined) set of coherent beliefs (Williams 1981c; Keil 2007, 142). People who intentionally lick the stones of St. Paul's cathedral in order to taste their color cannot be considered moral people because they lack the latter requirement. Their beliefs are utterly misled. People who (intentionally) book flights to Vietnam—because it has always been their dream to visit Asia—and book flights to Paraguay because they always wanted to visit South America—and then cancel both the next day exhibit stark unsteady behavior. If they continue doing this, doubts about their credibility as moral persons naturally arise and maybe friends of them will consider a therapist. They do lack the kind of coherence over time that we expect moral people to exhibit. Although these people meet in particular instances Davidson's requirements for agency (intentionality), they cannot be considered moral people.

It is not in the scope of the present text to explain, in more detail, what is meant with "coherent behavioral patterns," "lasting ground projects" and "coherent beliefs." However, it is important to mention that moral personhood requires more than being occasionally an agent of the Davidsonian type. Now, given the conditions mentioned above, it is natural to consider French's arguments for assuming that corporations' behavior can be described as intentional. This is the crucial pillar of the argument. Here, the corporates' internal decision structure (CID-Structure) comes into play. This is the instrument that makes it reasonable, according to French, to describe corporate activities as intentional actions despite diverging intentions of the individual members of the corporate body. The corporate's internal decision structure provides the basis for a complex description of the corporate behavior as intentional behavior. He writes:

> Every corporation has an internal decision structure. CID Structures have two elements of interest to us here: (1) an organizational or responsibility flow chart that delineates stations and levels within the corporate power structure and (2) corporate decision recognition rule(s) (usually embedded in something called "corporation policy"). The CID Structure is the personnel organization for the exercise of the corporation's power with respect to its ventures, and as such its primary function is to draw experience from various levels of the corporation into a decision-making and ratification process. When operative and properly activated, the CID Structure accomplishes a subordination and synthesis of the intentions and acts of various biological persons into a corporate decision. When viewed on another way, as already suggested, the CID Structure licenses the descriptive transformation of events, seen under another aspect as the acts of biological persons (those who occupy various stations on the organizational chart), to corporate acts by exposing the corporate character of those vents. A functioning CID Structure incorporates acts of biological persons. (French 1979, 212)

French proposes an example that I will briefly outline before I develop a critique. When there is a corporation—for example, the Gulf Oil Corporation—that consists of three executives being in charge of deciding whether the company joins a cartel, it is possible that each has its own individual reasons. It is, for example possible that one of them has been bribed to vote for joining the cartel, another one votes for joining because they have their own monetary interests, and the last one refers in his decision on the expert reports that several departments delivered to the executives. French writes: if the occupants "unanimously vote to do something and if doing that something is consistent, an instantiation or an implementation of general corporate policy and ceteris paribus, then the corporation has decided to do it for corporate reasons, the event is redescribable as 'the Gulf Oil Corporation did j for corporate reasons f.' (where j is "decided to join the cartel" and f is any reason [...] consistent with basic policy of Gulf Oil, e.g., increasing profits) or simply as 'Gulf Oil Corporation intentionally did j.'" (French 1979, 214) In this passage, we can immediately detect a first flaw in French's argument. By claiming that the decision for doing φ, such as joining the cartel, is an intentional behavior only if it is in line—if it is consistent, as French writes—with the company's overall policy such as increasing profit, French begs the question of intentionality. Clearly, if increasing profit is the intention of the corporation as presumed in French's analysis, and φ is an action that increases the corporations profit, then φ is consistent with the corporate's intention.

Now the question is: In which sense does increasing profit reflect the corporate's intention? Why should the goal "increasing profits" be approached as the *corporate's intention*? This question is crucial because the corporate policy and its overall agenda are also results of an internal decision-making process that synthesizes the goals and intentions of a *variety* of people, and these are often divergent. It must be assumed that at a certain stage in the corporate history the board of executives or other people in charge have decided that increasing profits with such and such technologies or products or under such and such conditions (like joining cartels or avoiding them) is part of the corporation's policy. But this is in itself a collective action for which it is uncertain whether it really represents the corporate's intentions. This is just the exact problem introduced at the beginning of our discussion. If French's proposal is supposed to have any value, it cannot merely presuppose that there is a corporate agenda that reflects the corporation's "real" intentions, and any action of the board of executives that is not consistent with such overall agendas is not the corporate's intentions. By doing so, French would already invest a notion of corporate intentionality, whose existence is in question here. What would be the solution to this problem?

The only possibility within French's framework is arguing that any decision made in agreement with the corporate's executives, or decisions that are the result of the CID structures, represent corporate intentions. Such decisions can result in processes that eventually harm the company, such as Ford's clinching to the Pinto which turned out to be an economical flop. They might also harm the company by putting thousands of jobs at risk, which is clearly not in the interest of the majority of the employees. Such decisions can also be unethical or, on the contrary, outstandingly praiseworthy. In summary, decisions that are not in line with the overall agenda and harm the corporation might not be considered rational, yet they *also* reflect the corporate intentions, if they are the result of a CID process. Excluding decisions that are not conducive to the overall agenda of the corporation, or which are harmful to the vast majority of employees for not being coherent with corporation's "actual" intentions beforehand, begs the question of corporate intentionality. If the CID Structure allows for such "irrational" decisions, then they must be considered as part of corporate's intentionality, too. Now, there is another troubling issue, which is best understood after recalling the general argument. The main line of reasoning in French's text can be simplified and reconstructed in the following manner:

> P(1): For every X: If X is a moral person, X is morally responsible.

> P(2): For every Y: Y is a moral person (and can, following from P(1), be held morally responsible) if and only if there is a description of Y's activities so, that whatever Y does makes true a sentence that says Y does it intentionally.

> P(3): Corporations' internal decision structure (CID) allows one to describe corporate activities as intentional behavior (hence, as actions).

> K: Therefore, corporations that employ a CID Structure can be held (morally) responsible.

As mentioned before, P(2) expresses a minimal notion of agency based on Davidsons theory. This is not yet a full-fledged account of moral personhood. We simplify this matter in the spirit of French's argument and assume that it is actually an account of moral personhood. Otherwise, his argument would definitely miss the target of proof. Between P(3) and the conclusion K, we also overlooked spelling out explicitly an intermediate step, which is that CID structures allow through the ascription of intentionality, also the ascription of moral personhood that can be inferred from P(2) and P(3). Only together with this hidden assumption, K follows (supposedly) from those premises. Now, I wish to direct the reader's attention to P(2) again. The Davidsonian account of agency requires that one can describe some events in a way that makes a sentence that says these things happened intentionally true. For the sake of argument, let us assume that by referring to the CID Structure of companies, French has provided a satisfactory account of how to *redescribe* the results of the complex decision making processes in corporations as their intentional behavior. Does this proposal fulfill Davidson's account?

I have pointed this out before: if we have a closer look at P(2), we see that Davidson combines two distinct conditions: first, that there is a description of a series of events that says it happened intentionally, and second, that this is a *true* description. We recall that we *can* describe the events in the bedroom as "Hamlet killing Polonius intentionally," however doing so does not provide a true description of these events. The only true description of these events in the bedroom says that Hamlet intentionally killed the person behind the arras. Thus, if and only if both conditions are met, we can properly speak of a Davidsonian type of agency. How do we normally know that someone did something intentionally?[84] In Hamlet's case, we know about his original intentions from his immediate confession about his mistaken assumption that the person behind the curtain was the King: "Thou wretched, rash, intruding fool, farewell! I took thee for thy better: take thy fortune."

As in this case, we often learn about intentions from a voluntary disclosure of agents. French mentions this when he speaks about answerability. People who are considered moral agents must in principle be able to share the grounds on which they acted. However, people can also keep silent or even lie about their actual intentions. When they are involved in a court trial, it can be difficult for the prosecutor to reconstruct whether the suspect acted in self-defense or with deliberate intentions. The suspect is in trouble if the attorney can prove that through the specific personal relation to the victim, they would have benefited from the person's death, that they apparently had a motif and if, furthermore, evidence reinforces this suggestion. Now, how about the company in our previously discussed Gulf Oil case? In the case of companies, the voluntary self-disclosure can only have two forms: Either, the corporate executives share their own personal intentions, which can differ vastly and do not even have to have any connection with the corporate fate. Or, they refer to the decision as being made in line with the corporate policy. Hence, regarding the overall goal of the company, which is (supposedly) increasing profit, the executives might argue that this was the reason *for the corporation* to join the cartel. The executives, thereby, do not put their own intentions forward but the corporate's intentions for joining the cartel.

Both strategies fail. The former fails because the intentions of the executives do not even have to be connected to the corporation's future and might diverge

---

84   Mackie points out that this is a factual question often hard to answer (Mackie 1990, 208).

substantially from each other. The latter one because—as argued before—it is an open question whether increasing profit (by joining the cartel) is really the corporation's intention. If increasing profit is part of the official corporate policy, it must have been established before through a process of collective decision-making in which the executives probably had the weightiest voices. Then, the question reappears: Why should we consider the overall agenda expressed in the official policy documents as an expression of the corporate's intentions? Especially, when we consider how starkly activities, such as joining a cartel or making profit, could affect the company's employees by putting their jobs at risk, we do not find a clear answer why in such cases the decision made by the board really reflects the corporate's intention, since the employees probably desire something entirely different. At this point, we go back to the question with which we started: Do corporations have their own intentions, or do they merely consist in aggregated and possibly diverging intentions of a number of individuals?

In short: French has proposed a way of *redescribing* corporate activities as intentional behavior referring to their origins in the CID Structure. He missed out to show that such redescription "really" pins the corporate's intentions down, which is the second condition that must be satisfied for being a Davidsonian agent. Given that self-disclosure about intentions is not a possible option regarding corporate intentionality, one must search for another method (like the criminal investigation) that shows that the redescription via the CID Structure reveals the intentional "nature" of the events. Alternatively emphasizing that corporate behavior has been in line with the overall corporate's policy, foists intentionality through the backdoor into the argument which is, therefore, not an acceptable reference point. Thus, while the CID Structure allows the redescription of corporate's intentionality, this way of approaching is not superior to non-intentional descriptions of those events.

Before ending the discussion of French's approach, it should be briefly outlined how my own position diverges from Patricia Werhane's and Edward Freeman's critique of French's argument. The authors argue: "[I]t turns out that, by appealing to the notion of intentionality as French uses it, corporate moral personhood makes little sense. Because of the various applications of intentional language, it is useful to call those phenomena that exhibit intentional behaviour, 'intentional systems'." (Werhane and Freeman 2010, 521) As a follow up on this passage, the authors outline a huge variety of events that can be described as intentional. I can describe the shutting down of my computer as: "My computer wants to annoy me today." We can describe manifold events as intentional processes; droughts as the wrath of god, storms as nature proving its exuberance, the looks of butterflies as nature showing off her beauty (see also chapter two). Werhane and Freeman, however, do no explicate their problem with such ubiquitous ascription. Is ubiquity itself the ground for the taking French's approach as absurd? French connects the ascription of intentionality to the CID Structures, and it is not implied in his argument that the attribution of intentionality to nature or other entities is acceptable.[85] It is the CID Structure alone that "licenses redescription of events as corporate." (French 1979, 214) In my previous discussion I pointed out that we often do not know whether a course of events is rightfully described as something that happened intentionally. Methods like the reconstructive investigations by public prosecutors only allow for approximations when answering questions of intentionality. It is the second condition of Davidson's account of agency that French fails to meet in his article: Namely to show that corporate behavior is *rightly* described as intentional behavior.

6.4   Responsibility Incorporated—Pettit's Account

More generally, one might question whether there is any value at all in attributing moral responsibility to corporations. For many years, such entities have been granted with a proper legal standing. They can be penalized for legal misconduct. So was the Union Carbide Corporation (UCC) from Connecticut sentenced to pay $470 million USD as compensation to the victims of the chemical accident in Bhopal, India in 1989. This penalty has been the final resolution of a thitherto-unique

---

85   Note that the post-phenomenological approach described in chapter two leads to the sort of ubiquitous ascription of intentionality of which Werhane and Freeman are sensitive about (Verbeek 2005). There, the unspecific condition for introducing new types of agency was an entities' impact on human action and behaviour, and I claimed that this applies to many (inanimate) things. In contrast to French's argument, this approach implies more obviously a reductio.

legal case of the Indian State against a multinational company. Although the amount of compensation for which the court settled has, for many of the approximately 200.000 victims, been dissatisfying and regarding the enormous suffering "absurdly inadequate," (Jasanoff 2016, 77) in a legal sense the body was at least not exempt from consequences and punishment. Since corporations, universities, and associations already have a legal standing and can be penalized by the law, why is it so important to also hold them morally responsible? Peter French says that "[…] to treat a corporation [merely] as an aggregate for any purposes is to fail to recognize the key logical differences between corporations and mobs." (French 1979, 209) In the previous section, it was mentioned that this is a failure of the existentialist account of collective responsibility. Their concept of a "collective" is so wide that some evident differences that impact individual responsibility fell out of the picture and were completely neglected.

Directed at my proposal in the introductory section of the present chapter, however, this objection is off the mark. There, I asserted that individual responsibility in collective settings is dependent on the alternatives open to the individual agent. This very broad proposal can easily account for the differences between mobs and corporations. The randomness and lack of organization precludes the possibility for individual agents to affect a mob's overall behavior. What can be done if being part of a group of demonstrators that is moving forward, chanting for their rights, and walking straight towards a bridge that cannot bear so many people? If you cannot escape the mob and no one is listening to your warnings, then there is nothing else you can do. This lack of possible impact exempts anyone in the same situation from co-responsibility.[86] In contrast, in many companies alternatives to shape overall strategies do exist. There are opportunities of participation and of influencing policies, especially for people who are employed in a higher position. These opportunities might entail certain risks for the agent, such as disapproval or contempt from associates, and they might be ineffective if one's in a minority. But most important for the present discussion is that there are means

---

86   Clearly, if you were part of the organisational committee that planned the march including the crossing of the bridge, you should have considered that it might collapse under the weight. The same applies to the architects and stress analysts of that bridge.

of *influence*, whatever the most rational decision will be in a particular case. The possibility of shaping collective behavior is in itself a constitutive factor for the co-responsibility of individual agents in such settings. This distinguishes mobs and corporations sufficiently regarding the individual responsibility of their members. But are there other reasons to promote a stronger notion of corporate responsibility?

In a well-acclaimed article, Philip Pettit has given a positive answer to this question supported by two distinct arguments. I will briefly discuss both of them before I conclude this chapter with my own suggestions. In the conclusive remarks of *Responsibility Incorporated* Pettit writes:

> What reason can there be, then, for persisting in the ascription of a corporate form of guilt to a people or a nation or to a body of believers? I think that doing so can have developmental rationale, to return to a thought from the beginning of the article. To refuse to ascribe collective responsibility to the grouping as a whole, on the grounds that the evil done was done entirely be the spokesbody, would be to miss the opportunity to put in place an incentive for members of the grouping as a whole to challenge what the spokesbody does, transforming the constitution under which they operate: making it into a constitution under which similar misdeeds should be less likely. By finding the grouping responsible, we make clear to members as a whole that unless they develop routines for keeping their government or episcopacy in check, then they will share in the corporate responsibility of the group; even if they have little or no enactor responsibility, they will have member responsibility for what was done. (Pettit 2007, 200)

Pettit understands the attribution of responsibility as a possible incentive that should not be overlooked. We could utilize the ascription of responsibility to motivate staff members in changing the corporate climate. The argument rests mainly on a descriptive premise. This premise is so the staff will be motivated when attributing moral responsibility to the company as a whole. It would be fatal for the argument if this were not the case. While, it might be true in some cases that we would activate tacit and critical resources within the company through responsibility ascription, it might just as well happen that the exact opposite results from such move. By attributing a general corporate responsibility people in the corporation, who had little influence on the overall agenda, might become discouraged and demoralized (Swierstra and Jelsma 2006, 314). As a result, they might give up doing anything at all in favor of a better and more ethical climate in the firm.[87] This result is, at least *prima facie*, as likely as the opposite course of events put forward by Pettit as a ground for an extended attribution of corporate responsibility and it has to be regarded as an undesirable consequence.

Moreover, the descriptive premise has another side to it. Who is the "we" in Pettit's argument? This matters insofar as it is after all highly unlikely that anything at all happens regarding corporate's moral behavior, if Pettit provides a theoretical argument for corporate responsibility and convinces a number of well-meaning philosophers to adopt his viewpoint. If change is demanded, it is (supposedly) rather futile when a philosopher leaves her role as external observer and starts wagging the finger at the company without having any connection to it or its staff members. If at all, it is arguably not the most effective measure to increase corporate conscience. If change is sought, there are probably better strategies— like creating financial incentives for moral behavior or rewarding whistleblowers, for instance, as Martins and Terblanche suggest (Martins and Terblanche 2003)— than extending responsibility ascriptions to the inflationary. Which measure is most effective is clearly an empirical question that differs from case to case, and it will likely depend on who employs these measures (Smith 2007, 478). This cannot be set our here in more detail. However, the most effective means to improve a corporation's moral behavior is not necessarily the attribution of responsibility to the corporation as such. Therefore, the incentive argument is a meager argument for attributing corporate responsibility.

There is, however, another argument in Pettit's text that is more profound. It is attached to a version of the Discursive Dilemma. The Discursive Dilemma introduces a situation in which a number of people make individual decisions that eventually result in a decision, which neither of the individuals intended or even

---

87   A similar but somewhat complementary argument aimed against the attribution of individual responsibility in economic processes has been proposed by Karl Homann (Homann 1993, 33). He claims that it would discourage economists and managers if they were approached with increasing individual responsibility for climate change and economic crises, which are caused not by single individuals (ibid., p. 33). Just like Pettit, Homann does not provide evidence that this will be the consequences of blaming them morally. For this reason both arguments are rather weak (Sand 2016, 344).

desired. As a result of the established decision-making arrangement, however, the outcome is the overall undesired decision which is legitimated through the whole spectrum of their individual choices by way of the specific decision-making arrangement:

> Imagine a commercial company that is owned and effectively run by its employees in the manner of a participatory organization. And now imagine that it faces an issue about whether to forgo a pay rise in order to spend the money thereby saved on introducing a workplace safety measure, say a guard against electrocution. Let us suppose that the employees have agreed to make the decision on the basis of the majority view on three separable questions; first, whether there is a serious danger of electrocution, by some agreed benchmark; second, whether the safety measure that a pay sacrifice would by is likely to be effective against the purported danger, again by an agreed benchmark; and third whether the pay sacrifice involves an intuitively bearable loss for individual members. If a majority thinks that the danger is sufficiently serious, the safety measure sufficiently effective, and the pay sacrifice sufficiently bearable, the pay sacrifice will go through: otherwise, it will not. (Pettit 2007, 197)

We face this situation. Now, after appropriate dialogue and deliberation, the employees make their decisions on the separated issues, whether there is serious danger involved, whether it is an effective measure, and whether the pay sacrifice means a bearable loss for individual members. All of them would individually oppose the pay sacrifice if they could vote for a pay sacrifice separately, but the procedure only allows for an indirect vote about the matter, and according to this, they produce the following (hypothetical) outcome:

Table 1: The Discursive Dilemma as originally presented in (Pettit 2007, 197)

	Serious danger?	Effective Measure?	Bearable Loss?	Pay Sacrifice?
A	Yes	Yes	No	No
B	Yes	No	Yes	No
C	No	Yes	Yes	No
Majority	Yes	Yes	Yes	**Yes**

This table shows that although none of them actually voted for a pay sacrifice, the decision-making arrangement to agree on pay sacrifice through majority decision will lead to a result in favor of the pay sacrifice. None of them wanted the pay sacrifice; they merely had their distinguished opinions about the other aspects of the decision. Now, Pettit asks: "But suppose now that some external parties have a complaint against the group, say, the spouses of the less-well-off workers, who think the pay sacrifice unfair. Whom, if anyone, can they hold responsible and blame for the line taken?" (ibid., p. 198) The answer he provides is that "the spouses can only blame the corporate group as a whole." (p. 198). This is an interesting case, since at first sight, none of them intended this outcome (which is a condition for being held responsible, as argued before) and, thus, it looks as if no one can be blamed for the pay sacrifice. The corporation, therefore, seems to be the proper placeholder to fill the void. But how strong is this argument really? There are three objections to this kind of reasoning.

First, as Matthew Braham and Martin van Hees have pointed out, we must recognize that, if the agents are according to the design of the situation *forced* to consider the three aspects of the decision separately (if prohibiting having certain thoughts is at all possible), then they are exempt from moral responsibility because they were not free from external compulsion, which is—as argued in chapter three—a necessary condition for being held responsible (Braham and van Hees 2013, 620). If, on the other hand, they were free to consider their decisions in the light of the probable decisions of the other people, they could have taken the probable negative result from each vote with their distinct preferences into account. Pettit merely says that they "vote as they judge." (p. 198) But that does not necessarily imply that one votes for the issues as if they were posed as distinct questions. From such a perspective the choice to be made is not how to decide regarding serious danger, but how one positions oneself regarding serious danger in the light of a possible pay sacrifice *depending* on what the others will do.[88] Through such reasoning, the voters might convince themselves that it might be better to be sure and vote against all three. If a majority of them comes to such a conclusion, there will be no pay sacrifice. Such decisions were also honest regarding the issues at stake. It would honestly reflect the voters' stance about—let us say—serious danger, because it is made in the light of serious danger for pay sacrifice. The previous considerations uncovered that the situation is less dilemmatic than the name suggests. There is really no trade-off for the voters if they chose to consider either the distinct aspects or the aspects in the light of the overall possible outcome. That is why they ought to do the latter. Now, it might still be the case that two of them accept the pay sacrifice and then vote accordingly, But then they are also really responsible and supposedly—assuming that a pay sacrifice is really bad even when considering the benefits of guarding against electrocution—blameworthy.

There is a more fundamental aspect about responsibility in such contexts that should be discussed. Remember that Pettit outlines this case as an instance of a wider problem to which corporate responsibility should provide the proper response. He is afraid that people can entertain such or similar decision-making arrangements to dissolve individual responsibility for group harm and thereby produce "responsibility voids." He argues that:

---

88   This is the reason from French for establishing his Extended Principle of Accountability (EPA): "Also he may be accountable for those nonoriginal and second effects that involve the actions of other persons that he obliquely or collaterally intended or was willing to occur as a result or under different descriptions of his actions." (French 1984, 134)

> [...] the failure to impose a regime of corporate responsibility can expose individuals to a perverse incentive. Let human beings operate outside such a regime, and they will be able to incorporate, so as to achieve a certain bad and self-serving effect, while arranging things so that none of them can be held fully responsible for what is done. This could be fixed so that the individuals are protected by excusing or exonerating considerations of the kind that we rehearsed earlier. (Pettit 2007, 196)

In the present quote, we can easily detect a shortcoming in Pettit's reasoning. If someone really establishes a regime that is sought to exempt him- or herself or other members of the group from moral responsibility, such an intentional act is in itself highly blameworthy and deserves most reprehension. Consider that the corporation in question had to choose between a decision-making procedure of the sort mentioned above and a procedure that allows for a direct vote. Such regime must have been established before. Now, responsibility, as outlined in the section on recklessness (chapter three), sometimes starts before the actual actions for which we want to articulate blame. The intention for deciding on a decision-making regime is *in itself* a decision for which one can be responsible. If the corporation has decided for the previously outlined structure, they have intentionality accepted that in certain cases—such as the one outlined above—the overall result of the process might not properly reflect the individual voters preferences and that a kind of "responsibility void" emerges. If they had good reasons to do so, for instance, because any other decision-making procedure suitable for the corporation, its members, and their particular business would have had worse effects, then they also really stand behind that choice and the results it might produce.[89] This is anal-

---

[89]   Ibo van de Poel notes pessimistically: "It seems hard to hold anyone responsible for the fact all decision procedures have flaws (as far as we know from the literature on the subject)." (van de Poel 2015b, 68) On the one hand, it is unlikely that all decision-making procedures are equally flawed. There are certainly better ways to arrive at an agreement than the one outlined by Pettit. Responsibility means choosing for those better alternatives. If, on the other hand, all decisional procedures are

ogous to the reckless driver who deliberately chooses not to care about his responsibility as the steersman of a dangerous vehicle who bears the responsibility for
accidents even if he had no direct intentions about them.[90] Previous intentions span
over events that were not directly intended but directly accepted as a probable
result of actions. This also applies to responsibility. Consider, for instance, that
there are a number of deficits in democratic decision-making. It is said to be slug-
gish and does not allow for the reaction to social problems as radically as we would
sometimes wish. These are certainly flaws in the design of democratic decision-
making proceedures, which we accept to prevent evil that is more serious. The
individuals that deliberately fostered democratic thought and established democ-
racies are co-responsible for this design, but they are not blameworthy because we
(supposedly) willingly accept such deficits.

Moreover, we could radicalize the situation described above. In my first objection, it was assumed that all people involved in the Discursive Dilemma could
have taken into account how the overall decision is affected by the others' votes,
and they could have oriented their particular votes accordingly. The "serious danger"-decision must be considered as a "serious danger *versus* possible pay sacrifice"-decision. Clearly, if we modify this basic situation by involving many, many
more people and a two (or three) digit number of separate decisions from which
the judgment for another ultimate decision is derived through a complex mathematical function, then it becomes impossible to take other voters' decisions into
account and act accordingly. We assume, furthermore, that the overall decision
derived from this procedure is not comprehensible. It is, for instance, not clear that
the overall result is so devastating that the voters might decide to vote "no" just to
avoid such collateral damage. Therefore, assume that it is not clear whether collateral damage is the overall result. Now, from my point of view, this case clearly

---

equally flawed, then the responsibility vanishes because the agent's lack reasonable
alternatives.

90   As outlined in chapter three, I presume that the setup of making a driver's license
demands that one acknowledge the responsibility that accompanies driving. That
does not mean that you contemplate the problem which kind of driver you want to
become as you would over a philosophical problem for month or years. Drawing
proportionate attention on certain subjects depending on their significance is also
an aspect of responsibility (see also section 5.5).

resembles the kind of "mob situation" described above with the example of the demonstrators walking towards the bridge. We have a number of people with diverging intentions, who cannot affect the overall choice, which are, (supposedly) also clueless about the upcoming danger. I believe that there is indeed no one to be blamed. By arguing that "the spouses can only blame the corporate group as a whole," (Pettit 2007, 198) Pettit suggests that whenever individual responsibility vanishes under the opaqueness of a complex collective activity, the group as a whole is a legitimate "responsibility placeholder." But this does not follow. Holding no one responsible and accepting the inevitableness and tragedy of certain catastrophes is also a creditable response. Pettit's argument rests on an unjustified *exclusive disjunction* that says that either individuals *or* the corporation as such must be addressees of responsibility and no other appropriate alternative exists.

6.5 Collective and Corporate Responsibility as Pragmatic Foreshortening

With respect to the arguments in the previous sections, it should be clear why I am not convinced that it is meaningful to establish a strong notion of corporate or collective responsibility or to ascribe agency to these entities. I think that the legal standing of corporations, associations, universities, and political parties is a sufficient tool to restore justice in cases of their misconduct. However, since I admitted before that we sometimes speak of corporate and collective responsibility *as if* these entities can function like agents and responsibility addressees of this practice must be either given up or justified. I think there are good reasons not to give it up entirely—in fact, when speaking of *their* misconduct, as I just did, I utilized a concept of agency and attributed it to such bodies. Thus, the attribution of responsibility to collectives such as corporations should be understood as a pragmatic foreshortening. This is justified because we often do not know who the people that are co-responsible for group harms really are, nor do we know which of their actions were conducive to the harm. Collective and corporate responsibility is a foreshortening that is justified through a lack of insight, information in the actual decision-making process and particular individual involvement. Before finishing this chapter, let me briefly explore this idea through an example previously mentioned.

Townsend Car Ferries Limited managed the Herald of Free Enterprise that operated in the English Channel, which sank close to the Belgic cost in 1984. The Herald tragedy is introduced as a case of vanished individual responsibility at the beginning of Philipp Pettit's article, which has been discussed above. The Townsend Car Ferries Limited company was held (legally) responsible for the sunken ferry Herald and for around two hundred deaths of crewmembers and passengers (some sources say that the exact number is still unknown). Although this does not appear in his conclusions, Pettit's argument hints at the reasonableness of blaming Townsend Limited *morally*. In order to establish his argument, Pettit cites from the official report the following interesting passage: "[T]he primary requirement of finding an individual who was liable... stood in the way of attaching any significance to the organizational sloppiness that had been found by the official inquiry." (Pettit 2007, 171) Given this description, the sinking of the *Herald of Free Enterprise* seems to introduce a case where no one can be held responsible. It looks as if the accident emerged from systematic sloppiness within the company. And yet, what Pettit cites in this passage from the official document licensed and presented by the Department of Transport for the court officials is only half of the story (Department of Transport 1987). In other passages of the document we can read: "The HERALD capsized because she went to sea with her inner and outer bow doors open. From the outset Mr. Mark Victor Stanley, who was the assistant bosun, has accepted that it was his duty to close the bow doors at the time of departure from Zeebrugge and that he failed to carry out this duty." (ibid., p. 8) Pettit takes the *Herald of Free Enterprise* tragedy as an instance of a responsibility void, where individual failure seems to be untraceable and, therefore, not accountable.

If we followed Pettit's reasoning, the company—Townsend Car Ferries Limited—as a whole becomes a reasonable placeholder and addressee for responsibility. But apparently what Pettit presents from the report is only part of the story. While "systemic sloppiness" prevailed in Townsend's company, the major source of the disaster is clearly outlined in the above-cited passage. Mark Victor Stanley's behavior is determined to be the crucial origin of the ship's capsizing. Hence, is the responsibility void that Pettit points out really such a substantial void? This is exactly what we should critically assess and eventually negate. Having a closer look at the case, we find a number of individual failures that allow the ascription of co-responsibility. This—as we shall see—makes corporate responsibility of

Townsend Limited an almost futile conceptual response. Now, the failure to shut the bow doors must have had a number of preconditions. We do not know under which conditions Mark Victor Stanley was hired. If his lack of professional integrity (through previous negligence) was well known, it might have been a mistake to hire him in the first place. The report does not mention anything about this. Furthermore, it must also be mentioned that, besides Mark Victor Stanley, the report also explicitly points out the clear mistakes that have been committed by both Captain David Lewry in his function as the leading Captain of the ship and his Chief Officer Leslie Sabel. Both were suspended after the disaster for a period of two years (Sabel) and respectively one year (Lewry) from their positions (p. 74). Besides the open bow door, one of the main factors causing the ship's sinking was the high velocity to which the ship accelerated immediately after the start:

> Both the model tests and the Pride experiment indicated clearly that at Combinator 6 the bow wave would be well up the bow doors, i.e. perhaps 2 m above the level of the top of the spade. The Court has concluded that on the evening of the 6th March Captain Lewry did not follow the practice, which he described, of restricting speed so that water did not come above the spade. The Court is satisfied that the rate of inflow of water was large and increased progressively as the ship dug the bow spade deeper into the water and decreased the freeboard forward. (p. 7)

The vehicle accelerated up to approximately 18 knots, although only 14 knots were acceptable considering the weather and sea conditions on that day. Considering this failure in good judgment and the lack of oversight and security checkup that Lewry carried as the leading Captain of the Herald, the report concludes with a statement that leaves no doubt about Captain Lewry's individual responsibility for the catastrophe:

> Captain Lewry was Master of the HERALD on the 6th March 1987. In that capacity he was responsible for the safety of his ship and every person on board. Captain Lewry took the HERALD to sea with the bow doors fully open, with the consequences which have been related. It follows that Captain Lewry must accept personal responsibility for the loss of his ship. (p. 12)

Here, it becomes obvious that there are a number of actions that were conjoined and together caused the ship to sink which can be clearly determined and regarding which one can also clearly determine who and how responsibilities have been neglected. Apparently, as the report informs us, the Herald was managed for some time under similar sloppy circumstances. It frequently happened that the ship started taking off with open bow doors, and it was accelerating up to risky speeds. Now, besides the particular failures on March 6, the report also established that this behavior was not unique in the corporate's or the Herald's history. This kind of sloppiness prevailed throughout the whole Townsend enterprise. The authors of the report trace the main origin for the systemic sloppiness in Townsend Limited that became obvious during the investigation of the Herald disaster back to the issue of time pressure that was put on the employees: "[…] the officers always felt under pressure to leave the berth immediately after the completion of loading." (p. 10) Time pressure here is not put forward as an excuse to exempt them from responsibility. As a leading Captain of a ship with extended cruising experience, who also took all educational measures required to hold such a position, must stand above the time pressure put on him, just like a car driver who is obliged to make a reasonable judgment about the weather conditions in which he is driving. If, for example, mist impairs his vision too strongly, then he is obliged to let the car rest, even if he will be delayed. Still, if time pressure cannot exempt the Captain and the other staff members from their individual failures, we might argue that it is the cause that many distinct instances of sloppy behavior accumulated. If Lewry had not been accelerating the ship up to 18 knots, the Herald might have made it safely to the next harbor even with open bow doors. Hence, here we might reasonably speak of a corporate failure: the prevailing time pressure that has been put on all employees of Townsend Limited has been part of the corporate behavior that cannot be reduced to individual failure and that contributed to the Herald catastrophe.

This might be where Pettit is right, and we cannot resist but to establish the attribution of corporate responsibility as a response to such systemic failure. Again, as I indicated above, we can employ the ordinary language of corporate responsibility here as a *farçon de parler*. This is pragmatically justified since we are not familiar with the internal decision structures, hierarchies, and individual decisions that led to the emergence of a system that ranked time efficiency higher than passengers' safety. When we are better informed about a case—such as the Herald's disaster just discussed—we can determine individual responsibility (Mark Victor Stanley's, Captain Lewry's and Chief Officer Sabel's behavior were most blameworthy). It is likely that, by also regarding the systemic sloppiness of Townsend Limited, we were able to pinpoint some individual failures, if we had more insight into the internal (decision-making) structures of the company. As argued at the outset of this chapter, we would then evaluate individual decisions in the light of the alternatives given to leading managers and stakeholders of Townsend and by assessing their foregoing reasoning. Here, the collective and corporative settings provide the framework within which individual decision making might be restrained or empowered by design. While this is taken into account in the assessment of individual action, it does not undermine individual agency. It can be mentioned at this point that sometimes collective responsibility is put forward as a rhetoric strategy to dissolve individual responsibility in contexts of collective actions. When it is aimed to trace the sources of collective harms people use to point out their embeddedness in institutional designs, and the accompanied constrains and impairments of freedom in decision-making. However, sometimes the very same people welcome rewards granted to the companies as results of certain achievements, which they willingly accept as if *they* contributed to the corporate or collective success. This asymmetry cannot be upheld. In collective settings, individual responsibility (as in general) prevails for both good and bad conduct and arouses respectively praise or blame. This responsibility cannot be fully dissolved under the opaqueness of collective decision-making.

Whether there are degrees of co-responsibility arising from the particular circumstances in which a decision has been made such as the position of the person on the hierarchical ladder of a company (which affects her range of alternatives), its (causal) influence on the resulting events or, more or less, stark negligence goes beyond the scope of the present text to discuss. For now, the reader should be reminded of the purpose of this chapter, which aimed at defending the notion of co-responsibility as a sufficient concept for dealing with cases of collective misconduct. This holds since legal liability is in place to *restore justice* in cases of corporate misconduct, and since speaking of corporate responsibility, which is—as Patricia Werhane rightly pointed out—widely adopted in ordinary language, can be explained as a pragmatic foreshortening. We happen to speak out of convenience about the companies' responsibility when we actually mean the conjunction of all those individual failures that contributed to an evil, about which we are often too little informed or too convenient to list in detail, just as we happen to anthropomorphize and ascribe intentionality to other things (like computers, the weather and animals), which we do not initially understand.

In any case, even if we cannot pinpoint any individual responsibilities, we should not seek conceptual shelter under a strong notion of corporate or collective responsibility. Some collective actions that result in disastrous events are tragedies without foregoing individual failures. Collective responsibility is no adequate reserve if we are unwilling to accept that there are cases where no one is to be blamed. In the next chapter, I will return to individual responsibility and virtue ethics as a possible moral standard according to which to judge agents and their responsibility.

7. The Virtues and Vices of Innovators

> *So much has been done, exclaimed the soul of*
> *Frankenstein – more, far more, will I achieve;*
> *treading in the steps already marked, I will pio-*
> *neer a new way, explore unknown powers, and un-*
> *fold to the world the deepest mysteries of creation.*
> *(Shelley 1989, 57)*

7.1 Introduction

If we wish to understand individual responsibility, we first have to deal with the preconditions of responsibility ascriptions such as the conditions for agency. This was outlined in length in chapter four. What has been neglected so far is not the question: "*When* is a man responsible?"—as Schlick asks—, but: "*What* is a man responsible for?" (Schlick 1962c) I argued that there is a number of general preconditions to be held responsible amongst which are being able to balance reasons, to act according to those reasons and to have a number of alternatives to choose from in a given situation (notions that are, as we saw, not as simple as they seem at first sight). Following up on this discussion, we must face another dimension of responsibility. As we saw before, responsibility is often discussed with regard to not only its preconditions but also regarding the *right conduct*. Clearly, responsibility also involves that some behavior is granted more positive responses than other behavior. In short, when we talk about responsibility we also ask whether someone deserves praise or blame for *something*, which might be consequences, intentions, attitudes, or conduct. Such questions about *right* conduct and moral character belong to the field of normative ethics and I will strengthen a virtue ethical theory of responsibility that has been prepared in my critique of consequentialist responsibility in chapter three in the following. Let me introduce this notion emphatically with a fictional case.

Imagine you work in a mediocre IT company. On your way to the monthly meeting, you bump into a colleague whom you only occasionally talk to. She is

© Springer Fachmedien Wiesbaden GmbH, part of Springer Nature 2018
M. Sand, *Futures, Visions, and Responsibility*, Technikzukünfte, Wissenschaft und Gesellschaft / Futures of Technology, Science and Society,
https://doi.org/10.1007/978-3-658-22684-8_7

ambitious, energetic, and usually busy in contributing to meetings and elaborating new ideas. Many of your colleagues consider her as a strong competitor. In today's meeting, she makes an even more confident impression than ever before and states that she is convinced that the company follows the wrong path with its new product line. Her critique is in line with objections you raised in the last team meeting. Eagerly she explains that she has a better and more innovative idea. She states to be certain that her idea directly reacts to current deficits in the market and that it is a smart and sustainable solution. The buzzwords she uses attract your attention and the preliminary outline of the project raises your interest and fuels your creativity. Her confidence is contagious and so is her determination. You are keen to deliberate the richness of her initial idea and after the meeting you are more than glad to hear that she appreciates having you on board for her project. This story could end with a successful innovation, a new technological pathway that fundamentally changes our society. It could also end with a failed start-up leaving you both with debts and unemployed.

There is something morally significant about both yours' and your fictive colleague's character that has been neglected in the discussion about responsibility in innovation processes. The passion and drive towards an alleged better future exhibited by your female colleague are characteristics that are, amongst others, relevant for innovation processes and these traits strengthen her persuasive power. Excellences of character and intelligence, such as her confidence, her rhetoric as well as her ambition make her a persuasive and, perhaps a successful entrepreneur; a person to whom your imaginary self, distinguished with similar features, is attracted to. The innovation landscape is filled with persons that possess some of these features. Peter Thiel, Ray Kurzweil, and Larry Page, for instance, are innovators who exhibit character traits like the ones previously mentioned. In the current debate on RRI the responsibility of innovators has been largely reduced to their causal role for social change as outlined in chapter three (Cabrera Trujillo, Laura Yenisa 2014). While the innovators' role for the development of new technologies must be acknowledged, their characters are rarely seen as a source for moral evaluation (noteworthy exceptions are (Pritchard 2001; Martin 2002)). The last chapter of this enquiry scrutinizes the virtues and vices of technological pioneers and what it means for them to innovate. It does so in three stages: First, the current approaches dealing with the responsibility of innovators and the place of responsibility in innovation processes will be critically discussed. Second, it will be demonstrated why the responsibility of a person cannot be exhausted in terms of his or her particular actions and their respective outcomes, but must be considered as expressions of her character. I will outline some benefits of Aristotle's ethical theory that makes reasonable a view that says that we are responsible for the kind of persons we are (Smith 2005, 242). Third, it will be shown that the language of virtue ethics can be fruitfully applied when assessing innovators' behavior. This will be achieved by scrutinizing two portrays of innovators of which one is fictitious (Victor Frankenstein) and one is real (Steve Jobs). The discussion will provide evidence for virtue ethics being a potent moral theory when discussing responsible innovation. Thereby this chapter provides a more nuanced understanding of responsibility that is not based on the "traditional" consequentialist model.

7.2 The Current Debate—Innovation, Visions, and Responsibility

Technological development influences our scientific endeavors, the way we live, culture, art, our understanding of aging and beauty, and the way we fight wars. Innovators can thus bring good and evil to our current and future society (see also chapter two). For decades in the philosophy of science, the multiple and often opaque ways in which engineers and innovators stand in relation to such events have been scrutinized. When questions of responsibility were addressed, the primary goal was to identify the causal chains between engineering activities and, for instance, climate change or other ecological or technological catastrophes like Chernobyl or Bhopal (Lenk and Ropohl 1993). Still today, innovators are exposed to criticism that links their daily work to the potential outcomes for society. Responsibility ascriptions have focused mainly on *causal* liability (Grinbaum and Groves 2013, 120). Especially actions that have negative consequences have been listed to attribute *ex post*—or as Ibo van de Poel calls it "backward-looking"-responsibility to innovators (van de Poel 2011, 37). However, the ways in which innovation practices are described and performed have become more advanced. Innovation processes have become globalized, non-linear, multi-layered and they involve numerous agents and stakeholders (Nowotny 1995; Tutton 2016; Urry 2016). The simplistic picture of the engineer bringing novelty into the world like a one-way street that leads via the producer to the consumer, is obviously outdated. Companies, governments, civil society organizations, innovators and consumers together form a variety of different innovation networks that have impact on these processes and together they create our technological future (Nowotny 2006; von Schomberg 2013). Innovation is, in other words, a process co-authored by many. This stands in stark contrast to the "traditional" picture of engineers as the primary or sole source for technological development. The current reality of innovation processes, including the already mentioned features, makes their governance particularly hard (Grunwald 2014b). The aim to shape the innovation process successfully is a common goal shared by many of the involved agents including the innovators themselves, politicians and their potential voters. These agents are permanently "negotiating the future," as Helga Nowotny described it (Nowotny 2006, 18). In the process of negotiating, technological futures and visions are getting increasingly important for all agents. Visions are used as a medium to communicate and homogenize expectations of stakeholders, policy makers and the wider public (Dierkes et al. 1992; Lösch 2006). Technological futures and visions are used as instruments to build up stable communities, to arouse interest and acquire the necessary approval and commitment from other agents (Geels and Smit 2000; Michael 2000). The fact that visions can be used to motivate and inculcate a team for a common goal became a common place in entrepreneurship as we can read in the textbook *Technology Ventures: From Idea to Enterprise*:

> Entrepreneurs need to create a shared vision or meaning for their venture. A dialogue of meaning and commitment will help bring a shared sense of urgency and importance for the venture. The vision can be written as a statement and verbally expressed as a story. The vision is used as a part of the business plan and described often to potential team members and investors. (Byers et al. 2011, 54)

It has been pointed out in detail by Patrick McCray that a technological vision has played an important role for the development of nanotechnology (McCray 2013). McCray's study of, amongst others, Eric Drexler's vision of nanotechnology, his community building activities and his influence on the development of the NNI is more than intriguing. Although the methods to steer innovation processes and the way of studying them have become more advanced, it should be clear by now that the rhetoric's and attribution of responsibility remained essentially the same. As I mentioned before, it is the "traditional" consequentialist and agent centred approach to responsibility that is still widely applied. Laura Cabrera, for instance, has passionately build on McCray's insights and enunciated her belief in the responsibility of visioneers for their potentially negative impacts on society (Cabrera Trujillo, Laura Yenisa 2014). She assumes that the extreme visions currently circulating in the field of many new and emerging technologies can cause radical changes in traditional structures and our social norms and values. They might influence research funding and policy (ibid., p. 205). Other authors joined her plea (Simakova and Coenen 2013; Ferrari and Marin 2014). In the following, I will briefly outline the deficiencies of the consequentialist model of responsibility before I propose a more suitable agent centred approach of responsibility.

There are several formal reasons for rejecting the consequentialist model of responsibility. To recall again this consequentialist idea is that we are mainly (or solely) responsible for the outcomes of our actions like innovating and constructing. Afterwards, when things went wrong, consequentialist responsibility implies being addressed with moral dispraise. This *ex post* feature of consequentialist responsibility is sometimes considered as a tool to "restore" what has been broken (Andre 1983, p. 205). It is understood as an instrument for criticisms, to influence and restrain certain innovation or engineering activities and eventually to improve engineer's moral behavior (Florman 1996, 40; Pritchard 2001, 391). While there are institutions that are devoted to the praise of the achievements of innovators (like the Nobel Prize), the responsibility discourse is mainly focused on (possible) failures of innovators and engineers. Overall, much more philosophical effort has been spent on developing conceptual frameworks to hold people responsible, than to express gratitude and praise for their work. This might be explained with the sanctioning rather than rewarding role of jurisprudence in our societies, a scheme that has been applied to moral judgments in general by, for instance, Moritz Schlick (Schlick 1962c). Though, the narrowing of the responsibility discourse on the negative side of the coin of innovation practices can easily be rectified by increasingly praising (and rewarding) innovators who contribute to social welfare. A further formal issue in the debate as it stands at the moment, is that attributing responsibility for consequences in highly complex systems requires us to trace back the origins of certain effects, which is often hard or even impossible (Swierstra and Jelsma 2006, 314; von Schomberg 2013). Even if such connections can be reconstructed afterwards, such an agent's responsibility largely depends on whether she could have known what she brings about beforehand and whether she had reasonable alternatives to choose from, both effecting (or in some cases undermining) her moral accountability (see chapgter three). Being involved in an accident, for instance, means that someone has caused an event that could not have been known before (else it has not been an accident but an act of negligence, as argued in chapter three). This suggests the difference between causal responsibility and moral accountability.

Hence, it was argued, for instance, by René von Schomberg that "an ethics focused on the intentions and/or consequences of actions of individuals is not appropriate for innovation." (von Schomberg 2013, 59) Schomberg also argues that we should implement ethics in the design of emerging technologies and establish structures and norms that make a responsible innovation processes possible instead of focusing on the blameworthiness of certain distinguished agents like, for instance, visioneers and their practices (ibid., pp. 63–70). Multi-stakeholder involvement and advanced deliberation methods become key factors in Schomberg's approach for structural responsibility (p. 67). Schomberg's ideas are valuable contributions to lessen the moral burden for individual agents in complex and opaque innovation processes and to search for a framework that enables responsive practices (DeGeorge 1991, 163; Swierstra and Jelsma 2006, 314). However, if we agree that individual agents lack the capacity to influence innovation processes significantly, the question arises who is responsible for the implementation of the many structural advancements that Schomberg proposes? His approach presupposes the existence of an agent or a group of agents who are capable of reacting to moral obligations and execute the proposed changes (Sand 2016, 345). Hence, those agents are suggested to have control over at least some parts of the innovation process. It can be assumed that policy makers are naturally the first to be consulted for this task (Grunwald 1999, 232). But if any agent has as little influence on the innovation process as Schomberg suggests, then those policy makers remain as impotent as any other group or community in implementing the proposed structural reforms. By rejecting the agent-centred approach to responsibility for innovation processes, Schomberg throws out the baby with the bathwater. My proposal is to harmonize the agent-centred and the structural approach to responsibility because they are not exclusive (Coeckelbergh 2006, 245; Blok 2017). The

first step in this direction is to enhance the currently reigning and simple consequentialist notion of individual responsibility. When we assess which kind of people innovators are and what it means to be an innovator for them, we get a better
understanding of individual responsibility for the innovation process. This is
where character traits enter the debate about innovating responsibly.

7.3 Crediting Virtue Ethics

In the previous section some formal problems of the agent-centred approach of
responsibility have been outlined. The objections raised affect the consequentialist
idea of individual responsibility. They are, however, not sufficient to dismiss the
notion of individual responsibility altogether. We should consider whether a different ethical standard is more appropriate for evaluating innovators. The Aristotelian framework that I will defend in the following cannot lessen the complexity
and opaqueness of innovations, but it has at least two advantages. First, virtue ethics emphasizes the inherent value of certain character traits and their value for
manoeuvring successfully and responsibly through the maze of innovation. Those
dispositions, some of which have been mentioned in the introduction of this paper,
are generally acknowledged as sources of good practices and they are irreducible
to particular achievements or failures of innovators. Second, it is the pleasure and
happiness resulting in acting those dispositions out, usually neglected by other
ethical theories, which is granted deserved attention from a virtue ethical perspective. Some ethical theories such as Kantianism recognize moral duties only in their
nature as being universal, but not as being conducive to the agents' well-being.
These theories are self-other-asymmetric, as Michael Slote argues (Slote 1992,
39–49). Neither Kantians nor Utilitarians ascribe any particular value to the motive
of living a happy life, which is a major shortcoming and results in misjudging
innovators appetite for change. Virtue ethics has been contested for many reasons,
which go beyond the scope of this enquiry. One of them, however, has received
considerable attention and should be briefly discussed. The argument states that it
is impossible or at least hard to derive any *concrete obligations* from virtue ethics
to solve problems in applied ethics. Transferred to the innovation context, it could
be argued that being at a loss about the right thing to do when facing a decision

that has unpredictable effects on society, is a typical facet of technological development. How should innovators act in such situations according to virtue ethics?
This critique has been put forward by Robert Louden (Louden 2007, 206). Louden
argues that virtue ethics deals only in a derivative manner with moral obligations,
which are demanded when moral predicaments are faced.

My response to this objection is twofold. First, it must be underlined that
Louden begs the question when assuming solving cases in applied ethics means
justifying what *ought to be done*. If this requirement is taken for granted, virtue
ethics is excluded as a source for moral knowledge e*x ante*. In contrast, as I will
show in more detail later on, there is definitely something to learn from an analysis
of virtues. Such knowledge does not necessarily imply what ought to be done, but
it is nevertheless an extension of our moral knowledge. It provides, for instance,
guidance and orientation for becoming a more creditable person and avoiding
making moral mistakes. Certain dispositions and capacities help to assess risks
properly and, thereby, enhance good decision-making. Specifically in the innovation context, such traits can become extremely powerful assets. One can make the
"mode" of virtue ethics more intelligible by considering, for instance, two different ways of teaching a carpenter apprentice. One way of teaching is telling him
exactly what to do next, such as cutting off this or that piece from the board or
adding the table legs here or there. The other way is going to a museum and showing him some masterpieces of the modern or ancient masters of his craft. This latter
way of pointing out greatness is also a way of guiding and learning a lesson from
examples, without determining exactly what to do (Burnyeat 1980, 72). Aristotle's
Nicomachean Ethics provides many examples of excellences in practical reasoning in a variety of context without claiming, however, that such reasoning is applicable to everyone in these situations (Wiggins 1976, 48).[91] In section four of the

---

91   Aristotle believed that ethics is a science essentially different from, for instance,
      mathematics or logic (1094b 12–28). According to Aristotle, every science allows
      for only the degree of precision that the nature of its particular subject permits.
      Ethics covers such a wide variety of choices and situations that "we must be satis-
      fied with a broad outline of the truth." (1094b 21) It seems to me that the huge

present article, a thorough discussion of the habits of two innovators will be provided that should serve an analogous purpose as illustrating excellence by pinpointing masterpieces in a museum.

Furthermore, the problem Louden raises is not even specifically a problem of virtue ethics.[92] It is rather a problem of all normative ethical theories when applied to wicked problems such as innovating responsibly, which is the very core of the matter of being a "wicked problem." Consider also some recent cases in the field of applied ethics that challenge indeed all existing ethical theories. For instance, the "Erlangen baby"-case, which was about a pregnant but brain-dead woman, whose bodily functions have been preserved to rescue the foetus, arose fundament dissent between proponents of feminist, Christian, Kantian, and other ethical theories (Singer 1994, 12). It is unlikely that any ethical theory provides a simple key to just deduce proper obligations in these and similar situations (Pritchard 2001, 394). The "Erlangen baby"-case is just one of the most striking examples. Hence, the derivation of obligations from any ethical theory is much more troublesome than Louden presumes when he considers this a disadvantage of virtue ethics. It is, furthermore, noteworthy that most of our everyday life decisions differ essentially from the "Erlangen baby"-case or those controversially discussed in applied ethics. When we are choosing fields of study, a profession, a lifestyle, or our social environment, the terms and concepts of deontology seem to be somehow displaced and virtue ethics unfolds a natural plausibility. Many of

---

variety of significant choices and situations that Aristotle mentions with which ethics is concerned, is often narrowed down to those cases that arouse most societal dissent. Abortion, stem cell therapy and euthanasia are topics that require reasonable arguments and which must sometimes be resolved with obligations of certain kinds. But these cases do not exhaust the variety of moral choices we make throughout our lives.

92   Consider, for instance, one of the most ambitious approaches of applying an ethical theory to a wide range of moral problems: Peter Singer's Practical Ethics Singer (Singer 2008). Singer's book is an outstanding example for clear analysis. The reasoning devotes itself to an impartial standpoint, which Singer considers as typical for classic utilitarianism. However, the arguments presented neither exhaust themselves in concepts associated with utilitarianism nor do their most interesting aspects lie in utility calculations—think, for instance, of his potentiality-argument (ibid., p. 153).

those more ordinary decisions are made in the light of what contributes to a good life. These decisions require us to act *aptly*, *thoughtfully*, and *clear-sightedly*. Whatever that means in particular, those traits are clearly virtuous and they can become significant for innovation. Let me just give you a brief example, before I elaborate this focusing on innovators; mastering one's temper is sometimes mandatory when aiming to consume sustainably because of the multiplicity of purchasing options that have to be considered (Gjerris et al. 2016). In general, the negative moral sentiments directed against someone who exhibits permanent *narrow-mindedness* or *arrogance* is real and does not concern his or her failure to concur with particular moral obligations, but the inability to live up to the standards of an aretaic morality. In the absence of any such obligations many people try to enhance their character to achieve a happy life and avoiding making moral mistakes. These reasons underline the major contribution of a virtue ethical perspective to our moral knowledge. To conclude, knowing what ought to be done in certain situations is an important but not a sufficient condition for a proper moral theory (Williams 2006; Slote 2010).

7.4   Character and Excellences

The inherent value of certain character traits and happiness are thus the pillars for my defence of virtue ethics. Let me connect those pillars with the theme of responsible innovation by first setting out my understanding of Aristotle's moral philosophy, which is—as I must admit right away—highly influenced by James Urmson's reading of the *Nicomachean Ethics*. I understand virtues as moral excellences that enable agents to do the right thing when combined with other intellectual excellences such as wisdom (Urmson 1999, 33). Virtues can be considered as character traits or dispositions (1103b 20–22). Being virtuous means being naturally inclined to do the right thing if such motivation is accompanied with the merits of the intellectual excellences. It is admirable how certain agents perform particular actions regularly with a lot of ease, as if it is their second nature (Burnyeat 1980, 74; Nagel 1991b). Agents who possess such a skill perform (good) actions without any inner resistance, just like Socrates who fearlessly and willingly accepted his death (Foot 1978, 4). Some character traits are admirable in themselves. Consider the following lively description of a case from Urmson:

> Let us suppose that Brown is a strong, healthy, extrovert, full of self-confidence. He is at a meeting where a course of action which he believes to be wrong is very popular with the majority; he speaks out against the policy and has no difficulty doing so. Let us suppose that Smith, a shy, retiring, hesitant person, is also at the meeting and also disapproves of the popular view. He can bring himself to speak out against it only be a great and very disagreeable effort of will. [...] But for Aristotle, Brown is the man who has excellence of character; he is the man who acts effortlessly and as he wants to act, without any internal friction. Aristotle is not making a hopelessly wrong judgment about moral virtue; he is raising a different sort of question. The excellent character is that which a man will have who lives the most eudaemon life, the most choiceworthy life. (Urmson 1999, 27)

In Urmson's example, Brown clearly has the disposition or inner appetite to speak up for his believes. Smith has our sympathies because he fought his inner resistance successfully down. Still, he is not the kind of person whom we would point out as an ideal when, for instance, educating children, though both Smith and Brown are successful in expressing their disagreement. "Whom of the two would you want your children to become?", is the hypothetical question Aristotle asks us to answer to understand the idea of excellence (Urmson 1999, 27). Now, the interesting question is whether there is a common disposition of innovators that might be admirable as such and how to approach it. Innovating is the creative process of finding solutions for particular problems (Florman 1996; Martin 2006). As such it requires creativity (Taylor 1987; Nozick 2006, 38). Innovators exhibit much more distinctively the intellectual merit of being able to creatively solving practical and theoretical problems than ordinary people do. Thus writes Mike Martin:

> Most creative work involves a high degree of purposeful activity and effort. Creative individuals value creativity, directly and highly, where creativity is understood as an advance in the domain of expertise of the fields they have chosen to work in. [...] They want to make discoveries, and they strive to be innovative. They organize and structure their lives around their commitments to trying to be creative, as illustrated by Elion, Salk, Nash, the Curies, and Feynman. [...] Creative individuals experience hope for progress, frustration at delays, joy in discovery, delight in confirmation of results, pride in achievement, curiosity about the simple and the complex, bafflement about anomalies, admiration (as well as envy) for others' achievements, disgust and contempt for shoddy work. These "rational passions" are as integral to creative work as are technical skills. (Martin 2006, 429)

According to Aristotle's distinction being made in passage 1103a 4–10 between moral and intellectual virtues, it seems reasonable to add creativity to the intellectual virtues. Creativity is best understood as standing in a row with other excellences of intelligence like wisdom, prudence and cleverness (1140a 25–1144a 24), because creativity itself does not provide us with a motivation. Clearly, the possession of any of these intellectual traits does not guarantee their use for the right purpose, but without them, the best purpose is *unarmed*. In book six, Aristotle informs us that the excellences of character are futile without the excellences of intelligence (1143b 19–25; 1144a 6–9). Aristotle outlines in this chapter that there is a huge variety of practical or intellectual virtues for finding the right means to certain ends and using them correctly (Urmson 1999, 84). The excellences of character such as, for instance, bravery, fortitude and equanimity provide the appetite towards certain ends. The intellectual excellences help us to take the right path to reach them. Urmson writes, reminiscing passage 1144b 9–14 of the *Nicomachean Ethics*:

> So how could we display a settled mean disposition to action, which is to be of good character, if we had not the wisdom to determine just what action is appropriate in the circumstances on each occasion? And how could we determine the precise form of plan and execute actions which we had no desire of any sort to perform? Without wisdom, excellences of character would be like a man groping in the dark and not knowing where to go; without the desires of an excellent character, wisdom would have nothing to do. (Urmson 1999, 84)

Mike Martin who has lucidly put forward a similar perspective as developed here, suggested that there is a difference between creativity and moral creativity (Martin 2006, 426). Martin is troubled by the idea that creativity can be used for immoral purposes like building advanced war technologies, for example. Therefore, he suggests that we should consider creativity as a moral trait only when it is used for the right purposes (ibid., p. 426). In the previously sketched framework, which is close to Aristotle's genuine theory, the issue can be resolved when thinking of a good person as possessing virtues of character and intelligence, which, as mentioned before, must appear together. To be attributed with a virtuous character requires one to pursue the right goals. Aristotle says that a man's "courage is a noble thing, so its end is of the same kind, because the nature of any given thing is determined by its end." (1115b 22–24; 1116a 13–16) Therefore, a man who is fortuitously fighting for a terror regime is not virtuous at all. However, she can possess independently from her character traits excellences of intelligence like creativity and wisdom. As a person, however, she will only be considered as good, when she uses these excellences to realize the intentions of her moral character: "Both the reasoning must be true and the desire right," summarizes Aristotle (1139a 24). This is the other aspect that is neglected in the traditional (consequentialist) responsibility debate. Only the combination of moral character and practical intelligence will contribute to her happiness. Aristotle's theory raises our awareness for the fact that activities like solving challenges of innovation, for instance, the production of sustainable energy sources with maximal efficiency or the development of affordable medicine, can constitute a person's *eudaemonia*, the good life or "living or faring well," as Urmson translated passage 1095a 16–20 of the *Nicomachean Ethics* (Urmson 1999, 11). When thinking of the life of a ballet dancer and asking for her legacy, we would do her great injustice when we reduce her professional career to the *sum of pleasure* that her particular performances caused in the audiences over the years (Martin 2002, 2007, 3; Wolf 2010, 2). This is, I guess, the perspective a consequentialist must have on the matter around which he or she will have located the dancer's main social responsibility. However, she might have chosen to become a ballet dancer because this activity gives meaning to her life, maybe because she wanted to devote herself and her bodily movement to pure aesthetics or she had the desire to master her body up to artistic perfection. Acting those desires out makes her acting even more realistic and touching. One might rightly argue that fusing her initial desire and the proper training will even contribute to her artistic mastership and success. It should be obvious that acquiring such artistic mastership often involves hard work and suffering. Being engaged in meaningful activities is not necessarily *pleasant* in an ordinary sense. As Susan Wolf writes with reference to the examples of writing a book, finishing a triathlon, caring for an ailing friend, and campaigning for a political candidate: "Many of the things that grip or engage us make us vulnerable to pain, disappointment, and stress." (Wolf 2010, 14) Technological innovators for some time in the nineteenth and first half of the twentieth century must have also experienced the inherent value of their choice of profession, as Samuel Florman reports:

> But engineers did, I believe, find their work thrilling in a deep-down, elemental
> way that we think of when the word existential is used today. They felt fulfilled

as men. They felt a part of the flow of history. They loved their work and believed
it was inherently good. (Florman 1996, 10)

Consider the case of Steve Jobs, whose biography will be discussed more thoroughly in the next section. His lifelong desire has been to build a minimalistic and
perfect technological devise, a device that combines aesthetics and high-end functionality in a perfect manner. Jobs' desire can hardly be subsumed or evaluated in
terms of what is morally obliged to do, when being an innovator or engineer.
Building (or in Jobs' case supervising the building) of such a device seems to be
an integral part of Jobs' idea of leading a good life (Nozick 2006, 39). The following statement underlines this desire:

> My passion has been to build an enduring company where people were motivated
> to make great products. Everything else was secondary. Sure, it was great to make
> a profit, because that was what allowed you to make great products. But the prod-
> uct, not the profit, were the motivation. (Isaacson 2011, 567)

Robert Sternberg reports with reference to a number of empirical studies that creative individuals "really love what they are doing and focus on the work rather
than the potential rewards." (Sternberg 2003, 108) The previous quote amplifies
this suggestion. Listening to Jobs' passionate words it is hard to imagine that anything else could have fulfilled him in the same way. The existence of such a deep
(existential) desire and its fulfilment as contribution to one's happiness has to be
taken into account when talking about innovator's responsibilities (Martin 2006,
428).[93] Recent studies on the motivation behind also show that many user-innova-
tors willingly accept the transaction costs involved in improving technologies by
own effort because "the [process of innovating] can produce learning and enjoy-
ment that is of high value to them." (Hippel 2005, 60) Conclusively, we should

---

93   Before Bernard Williams and Thomas Nagel were reviving the theme of moral luck
      as a problem for morality, Aristotle clearly pointed out that the execution of virtues
      is a necessary but not a sufficient condition to live a good life. Beauty, wealth and
      good health also contribute to happiness and they are clearly not under one's control
      (1099b 3–5; Urmson 1999, 14). The formation of virtues, however, is in one's con-
      trol and requires permanent training (1103a 14–26; 1114b 13–1115a 3).

extend the moral language currently applied to innovators and their enterprises with concepts such as admirableness and vice. I will apply such a wider focus in more detail in the next section.

7.5 The Virtues and Vices of Innovators

It is often argued that the usage of novels and science fiction stories can support the development of reflective capacities to deal with new and emerging technologies (Miller and Bennett 2008, 602; Grinbaum and Groves 2013, 136–139; Urry 2016, 114). Science fiction stories intertwine individual choices, fate and technology. They provide a resource to reflect technological futures through complex narratives of science, technology and society interrelations. Such a source for the rationalization of technological development is, for instance, Mary Shelley's *Frankenstein, Or the Modern Prometheus* (Miller and Bennett 2008). This remarkable horror novel continues to fascinate readers until today. I will discuss *Frankenstein* with its lively description of the moral character of Victor Frankenstein to give another example of how the language of virtue ethics can be applied to technological innovators. Victor Frankenstein, the chemist with a soft spot for natural philosophy, stands in the centre of the story. The description in the novel is comprehensive and intimate in displaying the motives and character of Victor Frankenstein and his scientific ambition. The narrative contains morally significant information that is often missing in analyses in which Victor's or other engineer's actions are decontextualized and assessed purely with regard to achievements and failures of their particular actions which might be considered as a sort of "conventionalist reduction." (Ulrich and Thielemann 1993, 879)[94] The vivid (self-) description of Victor Frankenstein arouses sentiments in light of the dramatic course

---

94 Similarly Martha Nussbaum argues: "But why not life itself? Why can't we investigate whatever we want to investigate by living ad reflecting on our lives? [...] One obvious answer was suggested already by Aristotle: we have never lived enough. Our experience is, without fiction, too confined and too parochial. Literature [and

of the story. It raises mixed feelings about Victor's guilt, which makes the story particularly fascinating. Both Frankenstein and his family are described as possessing extraordinary merits and virtues (Shelley 1989, 38).[95] Frankenstein in particular is praised by Robert Walton, the explorer who finds him on his way to the North Pole, for his noble character, eloquence and gentleness (ibid., p. 28). He considers himself as a caring family member with great affection towards his parents, his friend Henry Clerval and his beloved adopted sister Elizabeth (p. 42). His childhood was, in his own words, "exquisite" and "domestic." (pp. 44, 53) In contrast to Elizabeth who is "of a calmer and more concentrated disposition," indulged in muse and enjoying more often the beauty of nature, his curiosity and "thirst for knowledge" runs him into science (p. 42). While his temper remains "sometimes violent" and his "passions vehement", he sticks to science and starts to study philosophers of nature like Cornelius Agrippa and Albertus Magnus until he is acquainted with the science of chemistry through Professor Waldman at the University of Ingolstadt (p. 43). Waldman glorifies the achievements of modern chemists who "have indeed performed miracles [and] acquired new and almost unlimited powers" and finds an eager student in Victor (p. 57). Frankenstein thus becomes more and more devoted to his studies. While he makes rapid progress, contact to his family and friends becomes rare. He is increasingly dragged into the infinite realm of the natural science. His ardor for chemistry becomes unbound and his previous motive for conducting research to "banish disease from the human frame and render man invulnerable to any but a violent death," (p. 47) is replaced by the

---

biographies, p. 46] extends it, making us reflect and feel about what might otherwise be too distant for feeling." (Nussbaum 1992, 47) See also (Putnam 1981, 86).

95   Colin McGinn's reading of Frankenstein that focuses on the character of the "monster" is extremely stimulating. His way of justifying the value of literary analysis is in large parts congruent with mine: "Above all, questions of character assume far greater prominence when ethics is approached in this way, since fictional works are all about the interaction between character and conduct. The orthodox focus on moral norms and types of action will be an inadequate tool. To evaluate someone ethically you need to be able to analyse his or her character, and fiction still provides the best conceptual equipment for doing that (and probably always will). In fiction, character is the sine qua non. Character is to fiction what space and time are to physics."(McGinn 2000, 175)

motivation of becoming a master and father of a living organism (p. 65). Frankenstein carries out his scientific enterprise in a state of extreme devotion:

> These thoughts supported my spirits, while I pursued my undertaking with unre-
> mitting ardour. My cheek had grown pale with study, and my person had become
> emaciated with confinement. [...] One secret alone possessed was the hope to
> which I had dedicated myself; and the moon gazed on my midnight labours,
> while, with unrelaxed and breathless eagerness, I pursued nature to her hiding-
> places. [...] My limbs now tremble, and my eyes swim with the remembrance;
> but then a resistless and almost frantic impulse urged me forward; I seemed to
> have lost all soul or sensation but for this one pursuit. (Shelley 1989, 66)

Frankenstein pursues his scientific enterprise with dramatic eagerness. He becomes blind for possible needs of his creation and neglects the risk of his creation
affecting his social environment or society in general. Frankenstein sinks into a
state of trance, devoted to the purpose of scientific advancement. He happens to
believe that he acts in a void of social detachment. His character traits, his rampant
temperament, his *stubborn* devotion to his scientific curiosity make him *restless*
and egocentric. Retrospectively he admits that he "should not be altogether free
from blame. A human being in perfection ought always to preserve a calm and
peaceful mind and never allow passion or transitory desire to disturb his tranquil-
ity." (p. 67) He neglects his family and friends and misses to reflect the purpose of
his scientific endeavor. Victor Frankenstein's responsibility would have been to
master his temper, to widen his perspective and reflect on the potential social dimension of his creation. He does not bear responsibility for the dramatic and unpredictable events that follow (the persons, who cast out the monster are also partially responsible for that), but he is guilty of a lack of *moderation*.[96] Frankenstein's lack of resistance against his thirst for knowledge is a vice and the humus
on which the tragic story unfolds.

---

96   Aristotle's notion of σωφροσύνη is sometimes translated as placidity and some-
        times as moderation Aristoteles (Aristoteles 1995, 2003). Aristotle opposes
        σωφροσύνη to appetence, which is a vice of the irrational part of the soul (1117b,
        20).

Biographies often provide even more details about the character of the protagonist than some fictional stories like *Frankenstein*. In the biography of Steve Jobs, the "the ultimate icon of inventiveness, imagination, and sustained innovation," as he was called by biographer Walter Isaacson, we find a particularly interesting subject and heaps of data to analyse (Isaacson 2011, xxi). I do not want to suggest thereby that he stands, as Victor Frankenstein, at the beginning of a tragic course of (cruel) events for which he is morally blameworthy. However, moral evaluation, as I have mentioned before, concerns more than mere schemes of acts and consequences and of both Frankenstein and Jobs we have detailed descriptions of their characters and how they interacted with their respective environments over time. These descriptions reveal virtues and vices worth of being analysed. Steve Jobs was one of the founders of Apple, which is since 2011 the most valuable company of the world. From the early beginnings of Apple, Jobs was a person hard to get along with. Jobs has been allegedly manipulative and abusive with colleagues to successfully accomplish projects or businesses (ibid., p. 54). When, for instance, Nolan Bushnell, the founder of ATARI, asked him to create a single-player version of Pong he convinced his friend, the engineer Stephen Wozniak, to design the program keeping to him that Bushnell promised him a bonus for every chip they saved in the design. Years later, when Wozniak found out that Steve kept the bonus to him he was very disappointed. Isaacson quotes Wozniak saying: "I wish he had just been honest. If he had told me he needed the money, he should have known I would have just given it to him. He was a friend. You help your friends." (p. 53) Steve Jobs tended to evaluate ideas and proposals of his colleagues in extremes, as either being "shit," "crap," or using similarly dismissive attributes or overpraising them as "insanely great." (p. 106) This behavior was often impolite and sometimes bluntly degrading as, for instance, when he told his assistant Andy Cunningham that her suit would look disgusting before a press conference in New York in 1985 (p. 188). However, many of Jobs' former colleagues agreed that because of his radical assessments and harsh critique of their work accompanied with his intensity, Jobs made them perform much better than they would have expected from themselves. Striving for permanent improvement, simplicity, and perfection Steve Jobs often neglected good manners and empathy. From Nolan Bushnell he also adopted an entrepreneurial drive in pursuing these goals that made him being quite the opposite of placid. As Isaacson puts it:

> Unfortunately [Jobs'] Zen training never quite produced in him a Zen-like calm or inner serenity, and that too is part of his legacy. He was often tightly coiled and impatient, traits he made no effort to hide. Most people have a regulator between their mind and mouth that modulates their brutish sentiments and spikiest impulses. Not Jobs. He made a point of being brutally honest. "My job is to say when something sucks rather than sugarcoat it," he said. This made him charismatic and inspiring, yet also, to use the technical term, an asshole at times. (Isaacson 2011, 564)

This description sounds familiar. We can find traits in Jobs character that resemble Frankenstein's: Jobs lack of *inner serenity* that Aristotle considered as a virtues character trait. By eagerly pursuing a clear vision of a perfect technological innovation, he often neglected the obligations involved in friendships or social relationships in general. Just as Frankenstein loses *placidity* and thereby neglects his family for a while, Jobs neglects his social relationships (he abandons his daughter Lisa and betrays Stephen), he lacks empathy and compassion for his fellow men.

In general, it seems that innovators perceive, as mentioned in the fictional story at the beginning, the world as being full of issues with a lot of room for improvement. Innovators will not capitulate before they have found a way to solve them. The desire to improve often comes with high costs; persons are used as a means to an end, risks are neglected, and social relationships sacrificed. However, as I mentioned before, in both the case of Frankenstein and of Jobs their creativity to solve problems can be considered as an excellence of their characters.[97] Certainly, Frankenstein's success of creating a living being is an innovation that means a large step forward regarding the level of engineering or chemistry available in

---

97   It has often been asked whether it is required to be "an asshole at times," like Jobs, to be a successful innovator (Martin 2006, 430). This is an empirical question hard to answer. In the case of Jobs, it is clear that his idea of mingling aesthetics and functionality built the fundament of Apple's success. Jobs was extremely convincing and had an overwhelming sense for marketing. It is, however, also clear that his stubborn focus on end-to-end products led him to overestimate the value of the Macintosh computer. This focus almost ruined Apple in the midst of the 1980s. Also, after his successful return to Apple his personal struggles had negative effects on the climate in the company and made some of their best engineers walk off. From an evaluative point of view it is obvious that an asshole remains an asshole, no matter how succesfull (see section 5.6 and 8.2).

his time. Jobs' vision of the home computer as an affordable and user-friendly device through the usage of a mouse and graphical interfaces revolutionized the computer industry. In February 1982 the *Time* magazine considered him as having "practically singlehanded created the personal computer industry." (Taylor 1982) This might be an exaggeration, however, Jobs has rightly argued 30 years ago that computers should not be considered as mere machines that fulfil narrow purposes. He foresaw that they could become integral parts of human workspaces and lives, if they had an attractive and user-friendly design. The virtue of being able to create novelty, putting artifacts together that have not existed before, giving them a new shape or applying them in contexts in which they have not been used before is built on the disposition to regard the present as an object of transformation and possible improvement. Samuel Florman describes this attitude as dissatisfaction with the present and calls the inherent drive "to change [the world], to make of it something different" the existential state of engineers (Florman 1996, 120). In a similar vein puts Homer Barnett the essence of the "creative urge" in a nutshell:

> The individual who craves a change because of this incentive is "fed up" with some routine, repetitive activity. He is satiated and oppressed, but not because the distasteful situation is too laborious or because it fails to bring honor, pleasure or compensation. It is unpleasant just because it is dull and stultifying. No small measure of this feeling exists as a component of the creative urge just discussed, but it can function independently as a dominant drive. (Barnett 1953, 157)

Both passages remind us through lively description that such desire is indeed felt by innovators and engineers. This was one of the distinctive features of the fictive employee in the introduction of the present paper. Remember also what we said about Steve Jobs' desire to create a technological device that flourishes at the intersection of the humanities and engineering, a device that combines Bauhaus simplicity, functionality and user-friendliness. Jobs has developed an appetite for simplicity and clean design. His own house in Palo Alto barely entailed furniture because he was too meticulous about its design.

Desire for change is often neglected by ethicists for being partial or standing in opposition to the very idea of morality which is—as Kantian's claim—a system of categorical imperatives, detached from individual preferences and desires (Slote 1992). Though, the examples show that the dissatisfaction with preconfigured patterns and the contribution to one's happiness by changing them should not be downplayed (Hippel 2005, 61). This seems to be a neglected but frequent motive of innovators. In this respect, individual happiness and social change are closely related. Jobs creativity has, however, never been accompanied with a moral desire to produce technologies that contribute to social equality or similar ends. In contrast to his business competitor Bill Gates, Jobs never became a strong philanthropist (Isaacson 2011, 105). While Jobs' outstanding creativity appears to be a valuable trait, his ends looks *prima facie* as little more than the "shallow" desire for technological beauty (and money). Much worse is the motivational development of Victor Frankenstein whose initial impulse is to conquer human mortality and physical decay (Shelley 1989, 55). This initial purpose might at least benefit his fellow humans, but soon after this motivation takes a back seat and is replaced by a more selfish end:

> A new species would bless me as its creator and source; many happy and excellent natures would owe their being to me. No father could claim the gratitude of his child so completely as I should deserve theirs. (Shelley 1989, 65)

7.6   Innovation and Excellences

As argued at the beginning of this chapter and in other passages of this book: Technological pathways become more and more opaque. Therefore, many scholars notice a tension between innovation and responsibility. Responsibility should become, as it has been argued, a structural feature of innovation processes. Responsibility means the implementation of ethical values and bottom-up governance of innovation, according to those authors. However, this approach requires that agents be at liberty to act responsibly, meaning that they can exercise control over some parts of a large and complex system to implement these measures. Before innovation processes can be structurally responsible, the prior addressee of responsibility must be the individual agent. Building up on this idea the aim of this paper has been to develop a way of reasoning about individual responsibility that does not rest on the "traditional" consequentialist model. It has been argued that there is much more to evaluate about the morality of the professional life of tech-pioneers than their particular achievements as the two cases discussed in the previous chapter have shown. What we acknowledge about tech-pioneers, amongst other things, is their *eagerness*, *drive*, and *creativity* (Csikszentmihalyi 2004, 156). These character traits can be contagious and motivating. When they are used for the right purposes, they can be considered as virtues. Such traits are also worthy as tools to make good decisions and because they can contribute to a *meaningful life* (Wolf 1997, 210, 2010). More generally, we should insist on the positive value of, for instance, creativity to master innovation responsibly. Anyone who thinks about change can benefit from such a trait. In developing new technologies innovators have to manoeuver through a force field of conflicting norms and values: ethics, aesthetics, consumers' preferences, laws, corporative statutes, natural constraints (which might restrict the size and shape of technological devices), preferences of civil society organizations, fittingness to the existing technological infrastructure and so on. A virtuous innovator does not override those demands but takes them into account and searches for a solution that satisfies as many as possible. Creativity is an excellence that supports the finding of good pathways through this force field despite notorious epistemic uncertainty. Either this is done by transforming or conserving existing patterns (the latter is just a sort of selective transformation through omission, see chapter two). In the innovation context resistance against one's ideas is almost inevitable and, hence, *eagerness* and *patience* are required for success (Sternberg 2003, 108; Martin 2007, 30).

People who do not set their hopes on technological but social innovation also need imaginative capacities and the tenaciousness to stand behind their ideas. Technological pioneers who perceive the present as an imperfect state are fine examples how to successfully influence and creatively rearrange technological futures. The excellences emphasized do not ultimately erase the wickedness and complexity of responsible innovation but they help dealing with it. As I mentioned in section three, virtue ethical guidance is provided through analysing examples from literature or other sources that capture the complexity of a life lived. Those examples function as archetypes for orientation. Since a balanced in debt discussion of two persons' life courses was presented in the previous section we can ask: What can managers, entrepreneurs and innovators learn from this discussion? Just like the carpenter apprentice in the original example, managers can now orient themselves (positively and negatively) according to the presented discussion. Our discussion is so to speak an analogon to the teacher's method of visiting the museum with his apprentice, as an essential layer of Jobs' character—his appetite for change—has been uncovered. In their acclaimed book *Built to Last* Jim Collins and Jerry Porras argue that one pillar of lasting and innovative companies is their permanent dissatisfaction with their present way of doing business, even when they flourish. These actors permanently ask: "How can we do better tomorrow than we did today?" (Collins and Porras 2002, 185) Hence, the importance of Jobs' character trait is here supported by managerial studies and functions, therefore, as a positive example. The same can be said about his strong vision of building "a great device" which expresses a non-monetary purpose that exceeds the shallow idea of "maximizing shareholder wealth" often put forward as the core purpose of many ordinary companies (Collins and Porras 2002; Csikszentmihalyi 2004; Byers et al. 2011). The importance of such a shared vision has been pointed out already in the second section of the present chapter. On the other hand, we saw that Jobs harsh treatment of employees and friends has also been a reoccurring habit. According to Martins and Terblanche it is crucial for a company that aims at encouraging creativity to evaluate ideas fairly and treat mistakes as learning opportunities instead of blaming them as failed experiments (Martins and Terblanche 2003, 70–72). The clear deficits that Jobs exhibits in this regard can again provide *exemplary insight*, this time in a negative fashion. It is obvious that the presented reasoning exceeds the context of technological development. Imagining the vision of a more democratic society or a society that is more independent from technology requires this creativity as much as promoting and designing a successful technological innovation. However, it is obvious that my reasoning exceeds the context of technological development. Imagining the vision of a more democratic society or a society that is more independent from technology requires this creativity as much as promoting and designing a successful technological innovation.

Unlike moral principles as the categorical imperative, which is supposed to govern one's behavior from the moment one recognizes its authority and generality, virtuousness has to be trained lifelong. Neither is there a guarantee that one will eventually succeed in becoming virtuous nor that such training will definitely result in a happy life. External forces such as a poor economy or deteriorating health can conflict with people's happiness. This is one of the reasons why Aristotle believed that ethics, the study of making good choices and faring well, is a preliminary for politics (1181b 16–25). Politics is considered as the system of incentives and laws that will—if properly elaborated—contribute to the development of good traits and happiness by curtailing external threats and hindering their emergence. This is no less true of the institutional settings in which the innovators are embedded. The designs of such institutions affect without doubt the execution of innovators virtues while being—according to the previous arguments—best shaped by an excellent person. There is a reciprocal relation between the virtues of decision makers and the designs of the institutions they are embedded in (Blok et al. 2016). Martins and Terblanche mention a number of conditions that enhance creativity in companies such as rewarding risk affine behavior, encouraging communication, seeing failures as learning opportunities and fostering divergence among staff members (Martins and Terblanche, 2003). How exactly these are to be put into practice and to which extend those measures should apply (some failures are too devastating, to be regarded as learning opportunities) must again be *clear-sightedly* determined from case to case. This is where we are repositioned to the *agential side* of responsibility. Hence, it is reasonable to plead in the innovation context for both structural responsibility in the form of good governance and individual responsibility as the adoption of apt dispositions like creativity and tenaciousness. Both types of responsibility can co-exist and even strengthen each other to create a potent normative framework for responsible innovation.

8. A Vision for the Future of Responsibility

> *The thought of a world altogether devoid of music
> or literature or art is the thought of a world that is
> dark indeed, but if one dwells on it, the thought of
> a world lacking a single one of the fruits of crea-
> tive genius that our world actually possesses is a
> depressing one. (Taylor 1987, 685)*

8.1 Summary

In these final remarks, the most important theses of this book will be elicited and
the topics innovation, visions, and responsibility will be consolidated in order to
answer the research question raised in the introduction. In the following section I
will also draw attention to two aspects previously mentioned that require further
emphasis in these conclusions and that deserve to be thoroughly considered in future research: the symmetry of praise and blame and their application to innovators, and the challenges and advantages of democratizing innovation (Hippel 2005;
Jasanoff 2016). Let us first summarize the main arguments that I have provided.

In chapter two, it was argued that an ethics of innovation should be conceived
as a *humanist* ethics of innovation. This means that the focus of ethics of innovation is the (human) agents that are capable of reacting and adhering to normative
demands. I established my thesis in opposition to the idea that technological systems *determine* human behavior, as suggested by Peter-Paul Verbeek's post-phenomenological approach. Verbeek contests instrumentalism for neglecting the en-
tanglement between human agency and technological systems and assumes that
technologies "answer moral questions." (Verbeek 2011, 42) In contrast, I argued
that the usage of verbs such as "answering" foists intentionality into the description of technologies' function. This move pushes Verbeek's arguments down the
wrong slope. Technologies are instruments for human beings. Technology constrains or passively enhances what *people do* by constraining or enhancing *their*

© Springer Fachmedien Wiesbaden GmbH, part of Springer Nature 2018
M. Sand, *Futures, Visions, and Responsibility*, Technikzukünfte, Wissenschaft
und Gesellschaft / Futures of Technology, Science and Society,
https://doi.org/10.1007/978-3-658-22684-8_8

alternatives like natural environments and other contextual aspects of human agency (Bieri 2013, 46). This role is entirely *passive* and cannot suffice as a precondition to establish new forms of agency that are human-technology assemblies. It is more adequate to address the *rigidity* that technological systems such as the energy systems and other established infrastructures exhibit with the concept of "technological momentum." (Nye 2006, 55) The concept of technological momentum captures technology's resistance against immediate and fundamental transformation without presupposing that there is no room for incremental change or slow reform. It is human agents' inevitable responsibility to deliberate within certain contexts of action constrained by a variety of conditions (technological, social and legal). When understood properly those contexts can even enhance the governance of technology by utilizing the regularities in human-technology interaction. I have also justified why metaphors from evolutionary theory are insufficient to describe and explain technological change: Evolutionary theory utilizes concepts such as "random mutation," "selection," and "genes" that cannot be transferred to the context of technological development. Therefore, the evolutionary metaphor is so fundamentally stretched as to lose its *explanatory value* (Grunwald 2000b, 56). Furthermore, it was emphasized that innovations are ambivalent, which is crucial for dealing with it; if we *decide* to stand passive to the possibilities of change then we might as well miss out on opportunities for betterment—this is the charge of omission. My position opposes both technological pessimism and the unswerving belief in technology's progress. Developing a reasonable standpoint towards innovation that neglects neither the potential for positive nor for negative transformation is already a central part of adopting a responsible perspective on innovation.

Chapter three fathomed the perils and pearls of attributing responsibility in complex settings utilizing the exciting example of visioneering. I introduced visioneers as a type of social agent that receives increasing attention in STS and TA research (Sand and Schneider 2017). Visioneers' tools are not technologies or scientific methods but visions of technologies. Current large-scale technosciences expand the powers of individual actors. Their development also rests on visioneers' imaginative ressources, communication skills and their creativity to construct narratives of technological futures (Grunwald 2012c). They fathom the limits of what is technologically possible and create narratives to foster the interaction between diverse actors. Innovations are driven by such discourses about what is feasible and desirable. In these discourses visionary players approach interested groups with persuasive narratives of novelties by visualizing their concrete application or adoption by society and thereby stressing their potential. This was the case with nanotechnology, and there are currently many other technosciences in which visioneers outline pathways, arrange cooperation, and initiate new developments: Synthetic biology, big data, and in vitro meat are just a few prominent examples. Many scholars think that visions are an important entry point for research in STS and TA because the envisioned technologies themselves are far from maturation, and their consequences for society (especially in the long run) are more than uncertain (Grunwald 2014a, 2; Schick 2016, 230). In this book, I utilized the case of visioneering as a starting point to scrutinize the compartments of Pandora's Box of responsibility attributions in complex innovation journeys. It was emphasized that there is a difference between accidents and actions as intended behavior that is morally significant. Hamlet is responsible for his *intention* to kill the King of Denmark despite failing in his pursuit. This intentionality is *indirect* in cases of negligence and recklessness, which has also been briefly discussed. A person who *consciously* drives too fast *accepts* high risks and possible lethal consequences even if this person does not intend to harm someone. This analysis, however, pushes us away from evaluating particular *actions* to evaluating wider *attitudes* and *character traits* that concern, for instance, the acceptance of risks. Furthermore, I emphasized that responsible behavior has to be realized within contexts of (sometimes-reduced) alternatives. Responsibility means the deliberate choosing of the better pathways open to an agent; meaning those pathways that are preferable for the better reasons. This can mean the choosing of alternatives that look *prima facie* immoral, which become intelligible for bystanders when contextual information is taken into account (Bieri 2013, 110). Aristotle's example of the ship captain that throws the cargo overboard is pertinent and puts this idea *in nuce* (1110a 5–15).

Chapter four scrutinized the notion of agency. It was shown that intentionality, which has been emphasized as the cornerstone of responsible agency in chapter two, is often associated with the notion of intending and doing things rather than other things—with the ability to intend and act otherwise. What exactly does that mean? I outlined how determinism fixates the future by virtue of the laws of nature that conjoin events as causes and effects with necessity. By decomposing the integral parts of this worldview, which are the principles of causality, nomological causality, and universal determinism, we saw that there is little reason to buy in such a demanding theory. Given the *constructed* character of nomological regularities that are most familiar and utilized in technology, a worldview that suggests nature's fixedness in an even stronger manner is hardly plausible. Furthermore, by describing events as related to causes and effects, the theory of determinism fails to capture what sparks our interest when dealing with responsibility, namely people's actions. Actions lie *methodologically* on a different scale than causality. Actions are explained by the reasons that people had for *performing* them. It is an equivocation to treat the (physical) events that occur during actions as the *explandum* of action theory. An adequate action theory makes these events intelligible *as action results* when explaining (justifying) why people brought them about. Compatibilists often speak of a determination of actions by our motives (Hume) or character traits (Schlick). Since deciding often means opposing one's initial preferences or inclinations, these approaches do not adequately represent the phenomenon of agency (Campbell 1967b, 74). I stressed that we are not helpless victims of our character, but continuously shape and transform who we are. Of such persuasion are self-forming acts that cannot be explained by previous character traits because the act is a precipitation from a previous character trait or disposition. This is why psychological determinism explains—if at all—only behavior and not (reflected) decision-making. Many compatibilists like Moritz Schlick assume that *responding* to immoral acts with blame and punishment is a practice whose meaning sustains the possible discovery of determinism's truth. Schlick also argues that by determining "upon whom the motive must have acted," we can sort out people who are appropriate addressees of blame and punishment (Schlick 1962c, 153). This notion falsely suggests that we can entertain such an idea *without* libertarian presumptions. Other authors repeatedly challenged Schlick's theory of responsibility as opposing retributive intuitions. Retributivism says that punishment is legitimate only if it is proportionate to the evil of an action. Other functions such a deterrence and reformation are seen as unjust responses to wrongdoing. As a descriptive ethicist, Schlick does not (and must not) determine when the conditions of sanctioning or punishing behavior are appropriate: The charge that he allows in principle the decoupling of this function from actual wrongdoing is off target. It was suggested that there are ethical theories (like utilitarianism) that will more often demand the divergence of someone's responsibility and the responses to this person's behavior in the form of sanctions and punishment (Rachels 1993, 106). Furthermore, it was set out that it begs the question of the free will debate to presume that punishment is just only if people are *genuinely* free (Wolf 1990, 20). This leaves us with a more advanced understanding of agency, which is essential for any theory of responsibility.

In chapter five, I explored the interrelation between the contingent aspects of agency and how we take this into account in moral evaluation. Contingent events can occur before intended ends are reached (consequential moral luck), and they also precede what people do as the social and economic environments in which one grows up, and the character with which one is born (constitutional moral luck) (Nagel 1986, 1991b). Consequential moral luck emphasizes that the moral assessment of people, who behave *immorally,* but are fortunate enough to not cause any harm, diverts from assessment of people whose actions are interfered by accidental events, which they could not control. It seems that people who drive recklessly but do not kill a child are treated and evaluated differently than people who drive recklessly, but accidently do kill a child. This *prima facie* inconsistency in moral evaluation has been reinterpreted in chapter five. The first step in order to clarify our understanding of this problem is to acknowledge that there is *no* moral difference between lucky and unlucky persons. A lucky driver is someone who carelessly risks the lives of other people while an unlucky driver is merely a victim of circumstances beyond his control. The gap regarding their *blameworthiness* diminishes after consideration. Second, it is important to mention that there is a "prosaic" notion of responsibility that can be understood as *rectification* or restoring justice (Andre 1983, 205). This recalls an important dimension of the concept "blameworthiness" that has been pointed out in chapter four: There is a variety of functions that sanctions and punishment fulfill, which can be detached from moral evaluation. The legal charge of negligent homicide can be understood as fulfilling a *rectificatory function* (to comfort the victims of such accidents) and additionally *deterring* others from reckless driving. The logic behind these reactions overshadows and influences initial moral assessments. The distinction between the evaluation of a person and our reactions to them can be transferred to innovators. Grinbaum and Groves are certainly right to consider Victor Frankenstein as an unlucky innovator *par excellance* (Grinbaum and Groves 2013, 138). This, however, does not mean that we must withhold a moral judgment. On the contrary, we can provide a balanced evaluation of Frankenstein in *virtue ethical* terms. This is outlined in more detail in chapter seven. Following up on this assessment is the question whether it provides sufficient ground for a certain *reaction* to Frankenstein as, for example, in terms of blame or punishment. Frankenstein's companion Walton only pities him for losing his cousin, fiancé and friend, and this seems like an adequate response to his misery (Shelley 1989, 201). These events are too extraordinary to see the *function* of deterrence as a plausible ground to resent and expose Frankenstein publicly. As another form of contingency with allegedly paradoxical implications, we discussed constitutive moral luck. Constitutive moral luck seems to determine our character traits from birth and those traits are objects of moral evaluations too. Clearly, vanity, envy, and parsimony are intolerable character traits, even when people resist "by a monumental effort of will" to act on such inclinations (Nagel 1991b, 32). Instead of showing that such character traits are beyond our control, the arguments considered in chapter five *presumed* that this is the case. I argued before that people are able to train and ameliorate dispositions of character, which is also a central assumption of the virtue ethical framework developed in chapter seven. Thomas Nagel, whose exposé of the moral luck problem initiated the debate about moral luck, regards the different types of moral luck as instances of a wider problem: Circumstances and events that affect actions beyond agents' control push us to adopt an *external perspective* on ourselves that diminishes the familiar perspective of agency. Nagel believes it to be impossible to reconcile theses internal and external perspectives and describes the situation as a paradox. I have argued that irreconcilability of *perspectives* is not worrisome: Perspectives are no propositions. It would require much stronger arguments to show determinisms' *truth* instead of just exposing its content in a description of the external view on ourselves. Strawson's analysis of the moral sentiments, which has also been discussed in chapter five, provides a more nuanced picture of the interrelation between emotions and moral reactions (Strawson 1962). This, however, cannot undermine the persuasiveness of *incompatibilist* arguments such as those presented in chapter four. Although Strawson insists on the naturalness of the moral senti-ments and the impossibility of disposing them entirely, this is far from certain.

In chapter six, we discussed collective and corporative responsibility. As other human enterprises, visioneering is embedded in relatively regulated social environments: It is obviously not a solitary activity but performed at universities, in collaboration with publishers and companies, in public, at conferences, and in meetings. Thus, the activities of individual visioneers might be affected by the framework conditions that prevail in these settings. This raises problems of re-sponsibility: Does individual responsibility dissolve in such a collective settings? How much influence do single persons have in corporations? Can or should such bodies become responsibility addressees? Such questions were first systematically discussed after World War II. Existentialists such as Karl Jaspers and, later, Larry May argue that the belonging to a community that performs immoral acts can taint all of their members with a type of moral guilt if they do not (mentally) distance themselves from those acts (May 1991; Jaspers 2012). This is suggested to be true despite these people's inability to change anything about the community's behav-ior. It was argued that the existentialist account misses its target of proof: Their conclusion does not refer to a special form of *collective* responsibility. Instead, the argument emphasizes the value of not being *ignorant* about the misbehavior of other members of one's society or the misbehavior conducted in other societies—which designates an excellence of character. Peter French on the other hand, trans-fers the principle of attributing intentionality to corporate bodies according to their characteristic of having clearly defined rules and mechanisms in determining cor-porate policies (CID-Structures) (French 1979). For his argument, French utilizes a famous definition of action by Donald Davidson. Central to Davidson's defini-tion is the *proper* description of a process as intentional behavior in order to be considered an action (Davidson 1980b, 46). My discussion showed that French provides a description of corporate behavior as intentional behavior leaving the question of *properness* open. It is often unclear whether and, if so, which kind of intentions a person had to perform an act. This pinpoints to the dead end into which French's argument runs when recognizing just how many (also inanimate) pro-cesses can be (and are) described as intentional. Furthermore, Philipp Pettit brings two different arguments forward to treat corporations as responsibility addressees. First, he interprets the attribution of corporate responsibility as an incentive for their members to act more responsibly. Against this, it was argued that such an objective can easily backfire, and thus making a whole corporation responsible has the potential to dissociate employees from their employer and further *discourage* earnest behavior. Second, Pettit concludes that the most reasonable response to "responsibility voids" is to hold entire bodies responsible. This argument, however, is inconclusive. If an event is not attributable to anyone, the collective as a whole is not necessarily responsible either. Furthermore, it is noteworthy that it is highly irresponsible when collectives consciously establish decision-making procedures that can lead to "responsibility voids" if there are better alternatives at hand. If, however, there are no such alternative decision-procedures available, responsibility diminishes due to a lack of reasonable alternatives as argued in chapter three. In contrast to these approaches, I argued that the attribution of responsibility to such collective and corporative bodies could be understood as a *pragmatic foreshortening* when we are uninformed about the *initiators* of certain processes. In other words: We talk about companies *as if* they carry responsibility, but we in fact mean to attribute it to those people who could alone or together have made a difference (which depends on their position and level of influence) and who are often unknown.

Chapter seven discussed the virtues and vices of innovators bringing a number of different streams of thoughts from the previous argumentation together. First, the visionary origins of many innovations were emphasized once again: It is part of many companies' policies to think of technologies in longer terms and, thereby, give the company an identity that is strongly associated with such narrative (Collins and Porras 2002; Byers et al. 2011). Once again, I pointed out the irreducibility of agency in an ethics of innovation. Programs such as RRI require agents to employ strategies for structural reform in order to make innovations more transparent, inclusive, reciprocal, and open (von Schomberg 2013). If an innovation process ought to have these characteristics, *someone* has to shape such processes and he or she will be confronted with the ordinary challenges of complying with these demands of responsibility. Third, it was shown why ethical theories such as consequentialism and Kantianism are inadequate normative frameworks to assess innovators. The epistemic issues, our limited foreknowledge, and diminished control over the opaque innovation process have been amongst the arguments to reject these theories. Furthermore, it was outlined that, in today's most famous ethical theories the value of living a meaningful life plays only an instrumental role. Agents cannot plainly entertain a reduced self-conception as increasers of pleasure or devotees of duty. People are intrinsically motivated to foster social relations and pursuing individual ideals of the good life, and these motives cannot be classified as purely egoistic either (Wolf 2010, 4). Therefore, it is important to extend the variety of normative judgments with judgments about character. Virtue ethics is not uncontested, and one of the strongest objections to this moral theory is allegedly that it misses to provide of guidance for resolving issues of applied ethics (Louden 2007, 206). This argument exaggerates the quality of other ethical theories in providing such guidance. In contrast to theories that equalize what is morally obliged and good with those acts that are most conducive to the common good, virtue ethics provides orientation through the analysis of *examples* and by emphasizing them like "prize exhibits." In this manner, we first introduced the central concepts of Aristotelian virtue ethics: excellences of character and intelligence. While excellences of character like courage can clearly be used for immoral purposes, virtue ethicists since Aristotle are eager to emphasize that only the interplay of the right reasoning (excellences of intelligence) and the right desires (excellences of character) can constitute virtuous behavior (1139a 24)(Urmson 1999, 84). It was argued that creativity is amongst the excellences of intelligence and more prevalent amongst innovators. This is not to be understood as an empirical thesis, but rather as a conceptual analysis: A *paradigmatic* innovator is envisioned as a creative person. Furthermore, I pointed out that, for some people, producing novelty and being creatively engaged in developing new artifacts constitutes meaning in their lives (Gardner 1993; Florman 1996; Csikszentmihalyi 2004; Hippel 2005).

In order to get a more nuanced understanding of the virtues and vices of innovators and to provide an exemplary orientation of the kind mentioned before, two examples were discussed: Victor Frankenstein, protagonist of Mary Shelley's world famous novel, and Steve Jobs, who was the CEO of Apple which was the most valuable and allegedly most innovative company in the world at the time of his death in 2011. Literature and biographies provide detailed information about the *development* of a character over time and equips this information with details about the *context* in which a person lived his life. This is the ground material for a virtue ethical dissection (Nussbaum 1992, 23–29). Both Frankenstein and Jobs are, at times, excessively driven and rave about their innovations in the making. Under the spell of this extreme focus, both tend to lose their clear-sightedness and temper and these are amongst the vices of these two famous innovators. Such assessment stands for itself, despite the success (Jobs) or failure (Frankenstein) of a techno-logical endeavor. This resembles the conclusions drawn at the end of my chapter on moral luck. The interplay of the irreducibility of agency (individual responsi-bility) to improve innovation processes and the social environment (structural re-sponsibility) that can support and encourage virtuous behavior has been mentioned in the last section and shall be addressed once more in the following outlook.

8.2   The Symmetry of Praise and Blame

Do innovators and visioneers carry a special form of responsibility? This has been the initial research question of this research. We have seen that, in order to give rise to a judgment about the responsibility of innovators and visioneers, a number of preconditions need to be met. First, a person has to have the capacities to bal-ance reasons, to consider their actions in light of possible external (moral) de-mands and act according to such considerations to be morally responsible. This clearly applies to most innovators and visioneers: They are full-fledged moral per-sons. However, to be responsible means more than being part of the moral com-munity. It also means that something about a person's actions or dispositions at-tract other people's attention and provokes a normative *judgment*. Trivial actions, like putting on shoes in the morning, do not arouse such judgment even though they are also attributable to the agent and, therefore in a strict sense, something for which they are responsible (Smith 2005, 480). The judgments provoked are usu-ally negative; but also particularly admirable acts can trigger positive judgments and *responses*. In order to be reasonable these judgments must take a number of aspects into account: What was the actual result of an action, what was known about those results beforehand—about the likelihood of their occurrence—, has the agent aimed at producing such results or at least considered that they might occur (direct or indirect intentionality), what were the alternatives if there were any, what does *doing* this or that mean to this person, in which contexts was this or that done? These different questions and their specific weight have to be considered and enmeshed in a complex evaluation.[98] Answering many of those questions requires knowledge that is generated by the social sciences including psychology. When it comes to applying abstract philosophical theories about responsibility to concrete persons or actions, philosophy has to team up with the social sciences to inform each other about the normative and descriptive aspects that matter for responsibility judgments. This is clearly a sphere of action for future research (Grunwald 2017, 111–113). The "real" intentions of innovators and visioneers will often eventually remain opaque for observers and bystanders. Nevertheless, social scientists and psychologists can reconstruct a person's motives based on the quality of an action and background information about the context, and the person's biography and character. This supports providing an approximation of a person's intentions when this person herself is not disclosing them or when her honesty is in doubt.

Studies that specifically scrutinize the motives and intentions of visioneers do not exist. Nevertheless, Patrick McCray gives us detailed insights into the history of some famous technological visions of the late twentieth century and their promotors (McCray 2012, 2013). From these studies, we know that visioneers are different from "normal" innovators in their pursuit of more far-fetched, revolutionary futures, and their personal commitment in creating them by doing research

---

98   In a recent paper, Armin Grunwald writes: "[…] the creation and assignment of meaning by using technological futures can be interpreted in terms of its responsibility, just as any action can. This happens in the framework of consequentialism because the consequences of the assignment of meaning are made the object of responsibility considerations." And later: "The ethical aspect concerning responsibility does not emerge from the fact that the debated futures might become reality, but rather merely from the fact that they are used to create and assign meaning [own emphasis]." (Grunwald 2017, 107, 108) Clearly, visioneers often aim at shaping public debates and attract attention and this is true no matter what the consequences of these actions actually will be in the future. The intention to affect the innovation processes in the suggested ways underlie many visioneering actions, as Laura Cabrera and Armin Grunwald rightly point out (Cabrera Trujillo, Laura Yenisa 2014; Grunwald 2017, 103). What I doubt (and wanted to indicate in the passage above), is that such consequences are the only thing that matter for moral evaluation. Innvators would have to regard themselves as sort of "good-effect-machines" and such self-conception threatens meaning in life as the commitments to long-term projects and the maintenance of social relationships (Williams 1981c).

and engineering. In this regard, Eric Drexler—a prototypical visioneer according to McCray's definition—is much closer to Victor Frankenstein, who also aims at pushing the boundary of what is technologically feasible at his time, not only a little, but *much further*. Frankenstein wants to transform humankind's *nature* by defeating death. In contrast to Frankenstein, Drexler reflects (if only in a naïve way) upon the social consequences of Nanotechnology. His book, *The Engines of Creation*, explores the vision of Nanotechnology in much detail, and its possible implementation and diffusion in society (Drexler 1986). This shows Drexler's more thorough and comprehensive willingness to *self-transcend* his individual motives compared to Frankenstein. Frankenstein is at a certain point interested only in sustaining the intoxication from workflow, entirely neglecting his family, and the possible effects of his innovation on society. This *narcissism* is perfectly articulated when he exclaims that "[n]o father could claim the gratitude of his child so completely as I should deserve [the species' gratitude]." (Shelley 1989, 65) In contrast to Drexler and Frankenstein, Steve Jobs is a more typical innovator whose vision of a seamless technological device that combines aesthetic and functional simplicity is less revolutionary and disruptive. Most innovations are more similar to those of Jobs and can be characterized as minor, incremental changes of existing patterns (Danneels 2004; Hippel 2005, 21). These incremental changes are also suggested to be more commonly acknowledged and more readily accepted than fundamental transformations or even rejections of reigning designs (Sternberg 2003, xvii). Despite these differences, these innovators share the capacity to express their ideas of novel technologies and future pathways in narratives that further raise imaginative resources. This is strikingly exemplified by Jobs' wording of the "insanely great products," and his way of framing and staging them at release parties, Drexler's term "nanotechnology" with all related sub-concepts and umbrella terms like "nanobots," and also Frankenstein's announcement to solve the "mysteries of creation." They all verbalize and mediate a vision of the future. They also share the *willingness* to improve the present (that is at least also the case with Frankenstein whose initial idea is freeing mankind from the burden of mortality). The willingness to improve stems from a deep dissatisfaction with the present, which in Frankenstein's case originates from the early loss of his mother and in Job's case from an unsatisfied desire for perfect (in functional and aesthetic terms) technologies. Each of these innovators has had the *courage* to promote and pursuit his vision of the future, and they were *eager* and *driven* to realize these imaginaries. Eric Drexler pursued his personal idea of nanotechnology with such devotion that he eventually lost touch with the community he helped create and became a discard (Rip and Voß 2013, 47). He seemingly has not been *open-minded* enough to transform his initial vision in order to make it connective to emerging players' ideas. This underlines that innovating and visioneering are not seamless activities. These can be accompanied with solitude, mental instability and insecurity: Innovators try to push others to enter virgin soil, which can require a lot of *stamina* (Gardner 1993, 386). There can also be no doubt that these innovators display enormous *creativity* in solving problems. Note that this is not necessarily conjoined with dissatisfaction about the present. People can be dissatisfied with their own lives or the living conditions of other people without being motivated to change them—like the phlegmatic. On the other hand, they also can have such motivation without being able to *imagine* better solutions (Sternberg 2003, 107). Thus, creativity should be understood as an outstanding *intellectual excellence* that distinguishes innovators from simply dissatisfied or dissatisfied and lame people.

This analysis could be further advanced with more data, thereby leading to a more comprehensive study of the characters of Drexler, Frankenstein and Jobs. However, it is important to note that such *exemplary assessment* provides a nuanced picture of what innovators and visioneers *can be responsible for*. This wide understanding of responsibility is more appropriate than the "ordinary" understanding of responsibility as blameworthiness of an agent for damage or harm that succeeds his innovations. It utilizes normative terms originating from virtue ethical theory like "creativity," "temper," "open-mindedness," "stamina," "eagerness," "narcissism," "moderation," and "courage." Clearly, this list is not exhaustive. Future research has to discuss these virtues in more detail, who exhibits them and who could therefore function as good example to provide normative guidance. Furthermore, it is worthwhile to explore more virtues and vices and their prevalence in some professions and to see whether these belong to the intellectual excellences or the excellences of character as distinguished by Aristotle. This can be an important resource for *selecting role models* and understanding excellence in some professions.

Two aspects about this type of assessment should be emphasized. First, this kind of assessment sustains itself *despite* the success or failure of each of these innovators due to lucky circumstances. Steve Jobs becomes arguably the most successful of the three of them. However, we do not have to refrain from judging his sometimes violent temper as outrageous because of his economic success and fame. Frankenstein, on the other hand, is the least fortunate regarding the consequences of his innovation. Nevertheless, his creativity and eagerness for knowledge stand out as admirable traits. Such *judgment* must not necessarily be accompanied with a certain reaction (Smith 2005, 238). It is often suggested that (moral) responsibility ascriptions are publicly made *expressions* for compensation or reformation of the agent. For decades responsibility was seen as a conceptual response to mankind's growing technological power (Mitcham 1987; Jonas 2003). Engineers, scientists, and innovators were increasingly put in front of (fictive) public tribunals to justify their work and respond to social and moral demands. This way of (publicly) wagging the finger at relevant actors is a form of *responding* to their responsibility, to make them *aware* of their power and *sanction* them. I suggested that those kinds of responses must not necessarily follow responsibility judgements (Smith 2007, 472). While innovators like Jobs, Drexler, and Frankenstein are responsible for the kind of persons they are, this does not have to enforce any kind of reactions to them. Reactions that fulfill the functions of reforming the innovator, pushing him to act better in the future, or deterring others must also be considered in the light of moral and social requirements, like other actions (see chapter four). Responsibility *attributions* that only worsen a conscience that is—as in the case of Frankenstein—already miserable, and has allegedly no deterring effects whatsoever, can well be forborne (Smith 2007, 482). Furthermore, in deciding whether to blame or sanction someone we have to consider descriptive knowledge: this means, for instance, answering the question whether innovators are rather hampered in their willingness to strive for virgin soil if there is an inflationary amount of responsibility attributions. Many studies suggest that it is more *effective* to praise outstanding behavior rather than blaming reprehensible behavior in order to stimulate creative working environments—and effectiveness is one (not the ultimate) requirement to be considered, to make reactions reasonable (Martins and Terblanche 2003; Sternberg 2003). Other considerations concern someone's suitability to function as an idol. Steve Jobs has left memorable impressions on many people during his life. When *acknowledging* his work and his impact on the computer and smartphone industry that has changed the world by granting him

awards and honorary titles, we aim to give *something in return* and hope to encourage others to adopt Jobs' more marital traits.

This brings us to the second aspect of this assessment that is worth mentioning and closely related to the first: the presented exemplary assessment is *balanced*. It praises innovators for being courageous, adventurous, and creative. It is
not confined to blame. In philosophy of technology the topic of responsibility was
primarily approached as a response to the possible wrongdoing of engineers and
innovators, while wrongdoing was largely understood as the *causing of damage
and harm* (Höffe 1993, 21; Bayertz 1995, 5; van de Poel 2015a, 23). Responsibility
seemed to be the adequate conceptual response to the growing power to do harm
on the environment or extinguish life on earth. This move puts the complete engineering profession under suspicion. Hans Jonas' *The Imperative of Responsibility*
is paradigmatic in emphasizing the existential threat to mankind through human's
increased technological power and extended reach of action, and the need for a
new ethics and more responsibility based on this (Jonas 2003, 26). Up until now,
it is not easy to find literature on technological responsibility approaching the topic
from a more balanced point of view (exceptions often come from other disciplines
than philosophy (Florman 1996; Pritchard 2001, 393; Csikszentmihalyi 2004)). It
is important to leave this *asymmetrical* understanding of responsibility behind
(Watson 2004d, 283). As Angela Smith writes: "Merely claiming that a person is
responsible for something, therefore, does not by itself settle the question of what
appraisal, if any, should be made of the person on the basis of it." (Smith 2005,
266) Innovators and engineers have created a vast palette of technological "won-
ders" that enhanced living conditions (think of refrigerators, light bulbs, print
press, spectacles, sunscreen), increased life expectancy in large parts of the world
(antibiotics, industrial agriculture), and created artifacts of astonishing beauty;
most of us are in awe when standing in front of the Eiffel Tower or a space shuttle
(Lenk 1994, 41; Martin 2007, 2–5). The *praise* of such *achievements* should become a more central aspect of discussions about technological responsibility.
Growing technological power also implies the power to radically transform the
world for the better, which has also been a central aspect of our analysis on innovation's ambivalence in chapter two. Particularly when we want to *encourage* imitators, we should also focus on praiseworthy innovations and characters instead

of moralizing entire professions and, thereby, creating an unappealing environment of anxiety and (moral) restrain (Swierstra and Jelsma 2006). A striking passage in Jay Wallace's *Responsibility and the Moral Sentiments* reflects both on the previously mentioned *symmetry* as much as on the evaluation of *persons* despite their possible luck in consequences:

> Consider the mature artist's responsibility for a striking and successful work of art. In praising or admiring such a work, we do not just think of it as a successful production that happens to have been causally related to the artist who produced it. Rather, our praise and admiration reflect a kind of credit on its creator, opening the artist to direct assessment in virtue of the qualities reflected in the work. We can say that the artist is responsible for the work of art in a way we cannot say that a very young child is responsible for her finger paintings, even if the latter should turn out to be lovely in their way. (Wallace 1994, 52)

Wallace makes clear at the end of this quote that even children can produce lovely pieces of art. But we *credit* creative people for pushing themselves through periods of insecurity, for reinventing themselves permanently, and reflecting on previous failures, and continuing to refine previous approaches in sometimes long-lasting trial-and-error processes (Gardner 1993, 68; Hippel 2005, 63). In this, we see again how responsibility is ambiguous in combining two distinct aspects: judgments about people, and the interaction with them. The latter is guided by a practical interest and the former by a theoretical interest, both which are inevitably intertwined. By emphasizing the substantial role crediting and praising can play in the responsibility discourse, we also extended our view of the scope of possible *means* to improve human behavior.

A similar lack of attention has been paid to the *meaning* of innovating and engineering as *creative activities* (Nozick 2006, 39). The significance these activities can play in the lives of people is not fully understood and largely neglected in discussions about responsibility. As Mihaly Csikszentmihalyi writes: "In every historical period there have been individuals who care for more than their own profit, who find **fulfillment** in dedicating themselves to the advancement of the common good [own emphasis]." (Csikszentmihalyi 2004, 10) While there is a growing philosophical interest in the idea that there is meaning in life and its association with certain professions and activities (amateur sports, playing instruments, maintenance of friendships), my exploration of the concept in this book was far from exhaustive. Meaning in life is sometimes suggested to expand the dualist picture of human motivations categorizing them as being either egoistic or moral, which is often equalized with what is impartially best to do (Wolf 1997, 224, 2010, 6). Here, meaning in life seems to be at odds with such dualist conception: It is certainly not conducive to the common good to nourish a friendship, but is it therefore automatically selfish to do so? This would be a weird standpoint given how tough, dragging and painful it can be to become an outstanding ballet dancer—to recall the example from the last chapter. As Susan Wolf rightly points out:

> [...] someone whose life is fulfilling has no guarantee of being happy in the conventional sense of that term. Many of the things that grip or engage us make us vulnerable to pain, disappointment, and stress. Consider, for example, writing a book, training for a triathlon, campaigning for a political candidate, caring for an ailing friend. (Wolf 2010, 14)

Csikszentmihalyi's quote indicates that there are people who have found meaning in activities that were also conducive to the common good. However, what if those notions divert? What is understood here as meaning obviously differs from a summation of pleasant experiences: Learning to dance or playing the piano can at times be enormously tough and frustrating. My exposure to the idea of a meaning in life through activities like innovating have only scratched the surface of an iceberg that requires more attention in future research.

Finally, it should be mentioned that the exemplary innovators considered in the previous chapters share a number of features that doubt that they are interested in representing the needs of a majority of people, which may be affected by their innovations. This aspect seems indeed worrisome: All three men are white and have been born and raised in the Northern Hemisphere. They took advantage of higher education and shared exclusive access to information and existing technologies from the beginning of their lives (Sand and Schneider 2017). Seen from a global perspective, they clearly belong to a privileged minority. This reminds us, as Patrick McCray writes, that "[n]ot all futures are created equal." (McCray 2013, 17) In their visions of the technological future, the "typical" values that stem from being a member of a privileged class become obvious. The vision of radical life extension that is promoted by many of the most influential innovators from California's Silicon Valley is a striking example (Drexler 1986, 114; Kurzweil 1999, 2). Through this vision it becomes obvious just how strong individual ideals are, and the way to approach the future are rooted in the specific cultural background in which innovators reside. When looking at radical life extension as envisioned by many high-tech entrepreneurs as a technological fix, we recognize a certain blindness for the social reality and the origins of diverging life expectations on a global scale (Sand and Jongsma 2016a, 304). It can seem ironic that innovators in the Silicon Valley philosophize about radical life extension while there are vast differences in life expectation on a global scale as well as within societies of the Northern Hemisphere originating from *social inequalities* (Overall 2005). Social inequalities affect access to education and health systems; they affect life styles and job decisions, each having an enormous impact on well-being and life expectation. Thus, the vision of radical life extension by technological means as currently more and more embraced in some of the visioneering communities in North America fails to take crucial aspects of our social reality into account and deludes itself to solve a problem that actually has a variety of origins to which many non-technological, ready-at-hand solutions exist (Grunwald 2009, 160). Still, the investments made in these industries are real and anything but minor.[99] To give another example for the gap between "elitist futures" and social reality: It is well-known that in order to compensate the losses from expensive clinical trials pharmaceutical companies focus on developing drugs for more frequent and chronic diseases to "maximize the number of prescriptions" instead of doing research on less common conditions because they promise less financial returns (Dumit 2012, 6). Similarly, it is becoming increasingly difficult to establish one's own ideas of privacy and reasonable consumerism that is also able to compete with the visions of big digital oligarchs, (Apple, Google, Facebook) who plainly determine those issues top-down (Jasanoff 2016, 166). Given the power of such societal agents, one must acknowledge the obvious *limits of influence* ordinary people have on

---

[99]   The California Life Company (Calico), which is a daughter of Google's Alphabet Holding Inc., started out in 2013 to solve the "problem of aging" with a budget of $750 million USD (Various 2016).

future-making despite the growing interest of innovation communities in forms of maker spaces, fab labs, citizen science platforms, and open source software with low hierarchies in recent years (Hippel 2005; Hennen 2015). John Urry summarizes emphatically that the future is not exactly democratically made (see also (Jasanoff 2016, 255)):

> [...] it has been shown that they [technological futures] are indelibly bound up with the power of social actors to shape futures or even to "have" a future. People, places and organizations that do not have a future are physically or metaphorically pushed into the slow lane [...]. Indeed, we have seen how "powerful futures" are almost literally "owned" by private interests, rather than shared across members of a society. Certain futures are embedded within contemporary societies which bring them into being, being performative. Actors seek to perform or produce a future, and this can be realized as a self-fulfilling prophecy. Thinking through futures highlights something not articulated in much social science, which is that power should be viewed as significantly a matter of uneven future-making. (Urry 2016, 189)

These are sobering facts to say the least. Is this not enough evidence to demand more individual responsibility from visioneers and innovators? This question can be clearly affirmed within the limits that have been discussed in the previous chapters. We should remember that such limits could be supplemented with a policy of democratizing innovation. I discussed before that the *arenas* in which futures are negotiated are far from being carved in stone. Just like characters and actions, the framework conditions in which people innovate and make the future can be *transformed*, as well as the conditions that first lead to the values reappearing in the design choices of innovators and the policies that illegalize monopolies or other mechanisms that distort social markets. There is clearly room for improvement, and it is noteworthy that, in contrast to the responsibility attributed to individuals, such a type of *structural responsibility* is not faced with the challenge of providing the conditions to live a meaningful life (Frankena 1973, 53; Railton 1984, 161). As Thomas Nagel writes: "Institutions, unlike individuals, don't have their own life to live." (Nagel 1991a, 59) Institutions can decide to foster governance purely with reference to what is most conducive to social welfare, which would constitute a form of *institutional consequentialism*. However, such policies must also be legitimate, and, in order to be legitimate, they must represent the interests of their citizens (Nagel 1991a, 38). Structural responsibility has to meander within the limits of what is reasonable to demand from agents as citizens. One must be careful in order to avoid alienation amongst those agents that ought to support such policies. Some researchers encourage citizen participation in order to establish more widely shared and down-to-earth visions to counter elitist visions and, thereby, making innovation processes more inclusive, responsive, and broad in the scope of alternative pathways (von Schomberg 2013; Cabrera Trujillo, Laura Yenisa 2014; Ferrari and Marin 2014; von Schomberg 2015; Grunwald 2017). This may be attractive, but only for people who already have a basic idea or at least an intuition about a technology in the making. In contrast, many people prefer to live a private and solitary life and do not find appeal in the idea of altercating with emerging technologies or even developing their own visions of novel technologies that could challenge reigning, exclusive visionary paradigms like radical life extension by technological means (Bieri 2016, 33). This imbalance in interests will create a different type of imbalance: People who already have a basic interest stemming from an advanced level of education and skills are more likely to join innovation endeavors.[100] Thus, it is not easy to democratize innovation without *coercing* social milieus that do not genuinely employ an affinity to innovations. In the problem of the legitimacy of political institutions and policies, the challenge of diverging ideas of the good life and the fragmentation of values return on a higher level (Nagel 1991d, 5). Furthermore, establishing policies presumes thoughtful, reflected, and courageous decision-making or in short: Virtuous *agents*, which is an idea in line with the central thoughts of the humanist ethics of innovation as outlined in the previous chapters. With a sense of resignation, Sheila Jasanoff's writes at the end of her book, *The Ethics of Innovation*, "[i]nside the iron cages of large technological systems, subject to the tyranny of the assembly line and the production quota, emancipation and responsibility for the self may sound like cosmic jokes." (Jasanoff 2016, 262) Far from being a cosmic joke, responsibility for

---

100   Eric von Hippel reports that most participants in open source projects like Apache
       (a server software) share an academic background. Also, in sport communities like
       kitesurfing and mountain biking in which user-innovations are frequent people with
       an academic background are more likely to make adjustments—innovations—to
       their sports equipment (Hippel 2005, 99).

the self—agency—is the most important source of any theory of responsibility. Angela Smith writes in accordance with my reasoning: "[…] being held responsible is as much a privilege as it is a burden." (Smith 2005, 269) If human agents were not free to respond to normative demands, responsibilities' many preconditions and its scope could not even be discussed as we just did, and nature would take its course—but that is not a comprehensive picture. In this manner, this book can be understood as a vision to advance both our understanding *and* our dealing with responsibility in the future.

References

Alvarez M, Hyman J (1998) Agents and their actions. Philosophy 73, 219–245. doi: 10.1017/S0031819198000199

Amato I (1991) The apostle of nanotechnology. Science 254, 1310–1311. doi: 10.1126/science.254.5036.1310

Andre J (1983) Nagel, Williams, and Moral Luck. Analysis 43, 202–207. doi: 10.2307/3327571

Aristoteles (1995) Nikomachische Ethik. In: Bien G (ed) Philosophische Schriften, vol 3. Meiner, Hamburg

Aristoteles (2003) Nikomachische Ethik. Reclam, Stuttgart

Aristotle (2004) The Nicomachean Ethics. Penguin Classics. Penguin, London

Arnaldi S, Ferrari A, Magaudda P, Marin F (eds) (2014) Responsibility in Nanotechnology Development. The International Library of Ethics, Law and Technology, vol 13. Springer, Heidelberg, New York, Dordrecht

Ayer AJ (1952) Language, truth, and logic. Dover Publications, New York

Ayer AJ (2013) Freedom and Necessity. In: Shafer-Landau R (ed) Ethical theory: An anthology, 2nd edn. Wiley-Blackwell, Chichester, West Sussex, Malden, pp 317–321

Bacon F (1985) Of innovations. In: Pitcher J (ed) The essays. Penguin, London, pp 132–133

Bacon F (2011) Über Neuerungen. In: Schücking E, Schücking LL, Klein J (eds) Francis Bacon: Essays oder praktische und moralische Ratschläge. Reclam, Stuttgart, pp 82–83

Banse G (2012) Innovationskultur(en) - alter Wein in neuen oder neuer Wein in alten Schläuchen? In: Decker M, Grunwald A, Knapp M (eds) Der Systemblick auf Innovation: Technikfolgenabschätzung in der Technikgestaltung. Edition Sigma, Berlin, pp 41–50

Barnett HG (1953) Innovation: the basis of cultural change, 1st ed. McGraw-Hill Series in Sociology and Anthropology. McGraw-Hill, New York

Basalla G (1988) The evolution of technology. Cambridge History of Science. Cambridge University Press, Cambridge, New York

© Springer Fachmedien Wiesbaden GmbH, part of Springer Nature 2018
M. Sand, *Futures, Visions, and Responsibility*, Technikzukünfte, Wissenschaft und Gesellschaft / Futures of Technology, Science and Society,
https://doi.org/10.1007/978-3-658-22684-8

Bayertz K (1995) Eine kurze Geschichte der Herkunft der Verantwortung. In: Bayertz K (ed) Verantwortung: Prinzip oder Problem? Wissenschaftliche Buchgesellschaft, Darmstadt, pp 3–71

Bessant JR (2013) Innovation in the Twenty-First Century. In: Owen R, Bessant JR, Heintz M (eds) Responsible Innovation: Managing the responsible emergence of science and innovation in society. John Wiley & Sons, Chichester, UK, pp 2–25

Bieri P (2013) Das Handwerk der Freiheit: Über die Entdeckung des eigenen Willens, 11th edn. Fischer, Frankfurt am Main

Bieri P (2015) Eine Art zu leben: Über die Vielfalt menschlicher Würde, 2nd edn. Fischer, Frankfurt am Main

Bieri P (2016) Wie wollen wir leben?, 7th edn. dtv, München

Bijker WE, Hughes TP, Pinch TJ (1987) The social construction of technological systems: New directions in the sociology and history of technology. MIT Press, Cambridge, Mass

Blok V (2017) Bridging the Gap between Individual and Corporate Responsible Behaviour: Toward a Performative Concept of Corporate Codes. Philosophy of Management 16, 117–136. doi: 10.1007/s40926-016-0045-7

Blok V, Lemmens P (2015) The Emerging Concept of Responsible Innovation. Three Reasons Why It Is Questionable and Calls for a Radical Transformation of the Concept of Innovation. In: Koops B-J, Oosterlaken I, Romijn H, Swierstra T, van den Hoven J (eds) Responsible Innovation 2: Concepts, Approaches, and Applications. Springer, Heidelberg, New York, Dordrecht, pp 19–35

Blok V, Gremmen B, Wesselink R, Painter-Morland M (2016) Dealing with the Wicked Problem of Sustainability: The Role of Individual Virtuous Competence. Business and Professional Ethics Journal 34, 297–327. doi: 10.5840/bpej201621737

Bogner A, Decker M, Sotoudeh M (eds) (2015) Responsible Innovation: Neue Impulse für die Technikfolgenabschätzung?, 1st edn. Gesellschaft-Technik-Umwelt. Edition Sigma, Baden-Baden

Borup M, Brown N, Konrad K, van Lente H (2006) The sociology of expectations in science and technology. Technology Analysis & Strategic Management 18, 285–298. doi: 10.1080/09537320600777002

Braham M, van Hees M (2013) An Anatomy of Moral Responsibility. Mind 121, 601–634. doi: 10.1093/mind/fzs081

Bruce S (2008) Introduction. In: Bruce S (ed) Three early modern utopias: Utopia, New Atlantis and The isle of pine. Oxford University Press, Oxford, pp ix–lxi

Burnyeat MF (1980) Aristotle on Learning to be Good. In: Rorty AO (ed) Essays on Aristotle's ethics. University of California Press, Berkeley, pp 69–92

Byers T, Dorf RC, Nelson AJ (2011) Technology ventures: From idea to enterprise, 3rd ed. McGraw-Hill, New York

Cabrera Trujillo, Laura Yenisa (2014) Visioneering and the Role of Active Engagement and Assessment. Nanoethics 8, 201–206. doi: 10.1007/s11569-014-0199-5

Campbell CA (1951) Is 'Freewill' a pseudo-problem? Mind 60, 441–465

Campbell CA (1967a) In defense of free will. In: In defense of free will: With other philosophical essays. George Allen & Unwin Ltd, London, pp 35–55

Campbell CA (1967b) The psychology of effort of will. In: In defense of free will: With other philosophical essays. George Allen & Unwin Ltd, London, pp 56–77

Chalmers AF (1990) Science and its fabrication. University of Minnesota Press, Minneapolis

Chalmers AF (2007) Wege der Wissenschaft: Einführung in die Wissenschaftstheorie, 6th edn. Springer, Berlin, Heidelberg, New York

Chappell T (2009) Impartial Benevolence and Partial Love. In: Chappell T (ed) The problem of moral demandingness: New philosophical essays. Palgrave Macmillan, Basingstoke, pp 70–85

Chisholm RM (2007) Human Freedom and the Self. In: Kane R (ed) Free Will, 5th edn. Blackwell Publishing, Malden, pp 47–57

Coeckelbergh M (2006) Regulation or Responsibility?: Autonomy, Moral Imagination, and Engineering. Science, Technology & Human Values 31, 237–260

Coenen C (2011) Extreme Technikvisionen und die gesellschaftliche Verantwortung der Wissenschaft. In: Bartosch U, Litfin G, Braun R, Neuneck G (eds) Verantwortung von Wissenschaft und Forschung in einer globalisierten Welt: Forschen - Erkennen - Handeln. Lit Verlag, Berlin, Münster, pp 231–256

Collins JC, Porras JI (2002) Built to last: Successful habits of visionary companies. HarperBusiness, New York

Csikszentmihalyi M (2004) Good business: Leadership, flow, and the making of meaning. Penguin, New York

Danneels E (2004) Disruptive Technology Reconsidered: A Critique and Research Agenda. Journal of Product Innovation Management 21, 246–258. doi: 10.1111/j.0737-6782.2004.00076.x

Davidson D (1980a) Actions, Reasons and Causes. In: Essays on Actions and Events. Clarendon Press, Oxford, pp 3–19

Davidson D (1980b) Agency. In: Essays on Actions and Events. Clarendon Press, Oxford, pp 43–61

Davidson D (1980c) How is weakness of the will possible? In: Essays on Actions and Events. Clarendon Press, Oxford, pp 21–42

DeGeorge RT (1991) Ethical Responsibilities of Engineers in Large Organizations: The Pinto Case. In: May L, Hoffman S (eds) Collective responsibility: Five decades of debate in theoretical and applied ethics. Rowman & Littlefield, Lanham, Maryland, pp 151–166

Dennett DC (1990) Elbow room: The varities of free will worth wanting, 3rd edn. MIT Press, Cambridge, Mass.

Department of Transport (1987) Herald of Free Enterprise. Report of Court no. 8074. Formal investigation. Her Majesty's Stationery Office, London

Dickel S, Schrape J-F (2017) The Logic of Digital Utopianism. Nanoethics 11, 47–58. doi: 10.1007/s11569-017-0285-6

Dierkes M, Hoffmann U, Marz L (1992) Leitbild und Technik: Zur Entstehung und Steuerung technischer Innovationen. Edition Sigma, Berlin

Douglas HE (2003) The Moral Responsibilities of Scientists: Tensions between Autonomy and Responsibility. American Philosophical Quarterly 40, 59–68. doi: 10.2307/20010097

Drexler EK (1986) Engines of creation: The coming era of nanotechnology. Anchor Books, New York

Dumit J (2012) Drugs for life: How pharmaceutical companies define our health. Experimental Futures. Duke University Press, Durham, London

Ferrari A, Lösch A (2017) How Smart Grid Meets In Vitro Meat: On Visions as Socio-Epistemic Practices. Nanoethics 11, 75–91. doi: 10.1007/s11569-017-0282-9

Ferrari A, Marin F (2014) Responsibility and visions in the new and emerging technologies. In: Arnaldi S, Ferrari A, Magaudda P, Marin F (eds) Responsibility in Nanotechnology Development. Springer, Heidelberg, New York, Dordrecht, pp 21–36

Ferrari A, Coenen C, Grunwald A (2012) Visions and Ethics in Current Discourse on Human Enhancement. Nanoethics 6, 215–229. doi: 10.1007/S11569-012-0155-1

Fischer JM (2007) Compatibilism. In: Fischer JM, Kane R, Pereboom D, Vargas M (eds) Four views on free will. Blackwell Publishing, Malden, Oxford, pp 44–84

Florman SC (1996) The existential pleasures of engineering, 2nd ed. St. Martin's Griffin, New York

Foot P (1978) Virtues and Vices. In: Virtues and vices and other essays in moral philosophy. Blackwell Publishing, Oxford, pp 1–18

Frankena WK (1973) Ethics, 2nd edn. Prentice-Hall Foundations of Philosophy Series. Prentice-Hall, Englewood Cliffs, New Jersey

French PA (1979) The Corporation as a Moral Person. American Philosophical Quarterly 16, 207–215

French PA (1984) Collective and corporate responsibility. Columbia University Press, New York

Gardner H (1993) Creating minds: An anatomy of creativity seen through the lives of Freud, Einstein, Picasso, Stravinsky, Eliot, Graham, and Gandhi. Basic-Books, New York

Geels FW, Smit WA (2000) Failed technology futures: Pitfalls and lessons from a historical survey. Futures 32, 867–885. doi: 10.1016/S0016-3287(00)00036-7

Gjerris M, Gamborg C, Saxe H (2016) What to Buy?: On the Complexity of Being a Critical Consumer. J Agric Environ Ethics 29, 81–102. doi: 10.1007/s10806-015-9591-6

Godin B (2015) Innovation contested: The idea of innovation over the centuries. Routledge Studies in Social and Political Thought. Routledge, New York, Oxfordshire, England

Grinbaum A, Groves C (2013) What is "Responsible" about Responsible Innovation?: Understanding the Ethical Issues. In: Owen R, Bessant JR, Heintz M (eds) Responsible Innovation: Managing the responsible emergence of science and innovation in society. John Wiley & Sons, Chichester, UK, pp 119–142

Grunwald A (1999) Ethische Grenzen der Technik?: Reflexionen zum Verhältnis von Ethik und Praxis. In: Grunwald A (ed) Ethik in der Technikgestaltung: Praktische Relevanz und Legitimation. Springer, Berlin, Heidelberg, New York, pp 221–252

Grunwald A (2000a) Handeln und Planen. Neuzeit und Gegenwart. Fink, München

Grunwald A (2000b) Technik für die Gesellschaft von morgen: Möglichkeiten und Grenzen gesellschaftlicher Technikgestaltung. Gesellschaft-Technik-Umwelt. Campus, Frankfurt am Main

Grunwald A (2008) Auf dem Weg in eine nanotechnologische Zukunft: Philosophisch-ethische Fragen. Angewandte Ethik. Alber, Freiburg, München

Grunwald A (2009) Vision Assessment Supporting the Governance of Knowledge - the Case of Futuristic Nanotechnology. In: Bechmann G, Gorokhov V, Stehr N (eds) The social integration of science: Institutional and epistemological aspects of the transformation of knowledge in modern society. Edition Sigma, Berlin, pp 147–170

Grunwald A (2012a) Ende einer Illusion: Warum ökologisch korrekter Konsum uns nicht retten wird. oekom, München

Grunwald A (2012b) Synthetische Biologie als Naturwissenschaft mit technischer Ausrichtung: Plädoyer für eine „Hermeneutische Technikfolgenabschätzung". Technikfolgenabschätzung - Theorie und Praxis 21, 10–15

Grunwald A (2012c) Technikzukünfte als Medium von Zukunftsdebatten und Technikgestaltung. Karlsruher Studien Technik und Kultur, vol 6. KIT Scientific Publishing, Karlsruhe, Hannover

Grunwald A (2013) Techno-visionary Sciences: Challenges to Policy Advice. Science, Technology & Innovation Studies 9, 21–38

Grunwald A (2014a) Modes of orientation provided by futures studies: making sense of diversity and divergence. European Journal of Futures Research 2. doi: 10.1007/s40309-013-0030-5

Grunwald A (2014b) The hermeneutic side of responsible research and innovation. Journal of Responsible Innovation 1, 274–291. doi: 10.1080/23299460.2014.968437

Grunwald A (2015) Synthetic Biology as Technoscience and the EEE Concept of Responsibility. In: Giese B, Pade C, Wigger H, Gleich A von (eds) Synthetic Biology. Springer, Heidelberg, New York, Dordrecht, pp 249–265

Grunwald A (2017) Assigning meaning to NEST by technology futures: extended responsibility of technology assessment in RRI. Journal of Responsible Innovation 4, 100–117. doi: 10.1080/23299460.2017.1360719

Gutmann M, Weingarten M (1998) Überlegungen zu Innovation und Entwicklung. Technikfolgenabschätzung - Theorie und Praxis 7, 11–19

Hampe M (2007) Eine kleine Geschichte des Naturgesetzbegriffs, 1st edn. Suhrkamp Taschenbuch Wissenschaft. Suhrkamp, Frankfurt am Main

Harman G (1977) The nature of morality: An introduction to ethics. Oxford University Press, New York

Hartmann D (1996) Kulturalistische Handlungstheorie. In: Hartmann D, Janich P (eds) Methodischer Kulturalismus: Zwischen Naturalismus und Postmoderne, 1st edn. Suhrkamp, Frankfurt am Main, pp 70–114

Hartmann D (2005) Willensfreiheit und die Autonomie der Kulturwissenschaften. http://www.jp.philo.at/texte/HartmannD1.pdf. Accessed 28 March 2018

Hartmann D, Janich P (1996) Methodischer Kulturalismus. In: Hartmann D, Janich P (eds) Methodischer Kulturalismus: Zwischen Naturalismus und Postmoderne, 1st edn. Suhrkamp, Frankfurt am Main, pp 9–69

Hennen L (2015) „Public Engagement" in Forschungs- und Innovationsprozessen. In: Bogner A, Decker M, Sotoudeh M (eds) Responsible Innovation: Neue Impulse für die Technikfolgenabschätzung?, 1st edn. Edition Sigma, Baden-Baden, pp 91–100

Herodotus (1958) Die Bücher der Geschichte: Auswahl III. Reclams Universal-Bibliothek, 2206/07. Reclam, Stuttgart

Hippel E von (2005) Democratizing innovation. MIT Press, Cambridge, Mass.

Hoefer C (2016) Causal Determinism. In: Zalta EN (ed) The Stanford Encyclopedia of Philosophy, Spring 2016. Metaphysics Research Lab, Stanford University

Höffe O (1993) Moral als Preis der Moderne: Ein Versuch über Wissenschaft, Technik und Umwelt, 1st edn. Suhrkamp Taschenbuch Wissenschaft. Suhrkamp, Frankfurt am Main

Hölscher L (1999) Die Entdeckung der Zukunft. Fischer Europäische Geschichte. Fischer, Frankfurt am Main

Homann K (1993) Wirtschaftsethik: Die Funktion der Moral in der modernen Wirtschaft. In: Wieland J (ed) Wirtschaftsethik und Theorie der Gesellschaft, 1st edn. Suhrkamp, Frankfurt am Main, pp 32–53

Honderich T (1990) Conservatism. Penguin, London

Hudson H (1993) Collective Responsibility and Moral Vegetarianism. Journal of Social Philosophy 14, 89–104

Hughes TP (2006) Culture and Innovation. In: Nowotny H (ed) Cultures of technology and the quest for innovation. Berghahn Books, New York, pp 27–38

Hume D (1975) Enquries concerning human understanding and concerning the principles of morals, 3rd edn. Clarendon Press, Oxford

Hume D (2003) A treatise of human nature. Dover Philosophical Classics. Dover Publications, Mineola, N.Y.

Isaacson W (2011) Steve Jobs. Simon & Schuster, New York

Isaacson W (2014) The innovators: How a group of inventors, hackers, geniuses, and geeks created the digital revolution. Simon & Schuster, New York

Janich P (1981) Natur und Handlung: Über die methodischen Grundlagen naturwissenschaftlicher Erfahrung. In: Schwemmer O (ed) Vernunft, Handlung und Erfahrung: Über die Grundlagen und Ziele der Wissenschaften. Beck, München, pp 69–84

Janich P (2009) Kein neues Menschenbild: Zur Sprache der Hirnforschung, 1st edn. Edition Unseld, vol 21. Suhrkamp, Frankfurt am Main

Jasanoff S (2016) The ethics of invention: Technology and the human future, 1st edn. The Norton Global Ethics Series. W.W. Norton & Company, New York, London

Jaspers K (2012) Die Schuldfrage: Von der politischen Haftung Deutschlands. Piper, München, Zürich

Joas H (2012) Die Kreativität des Handelns, 4th edn. Suhrkamp Taschenbuch Wissenschaft, vol 1248. Suhrkamp, Frankfurt am Main

Jonas H (2003) Das Prinzip Verantwortung: Versuch einer Ethik für die technologische Zivilisation, 1st edn. Suhrkamp Taschenbuch Wissenschaft, vol 3492. Suhrkamp, Frankfurt am Main

Kaiser M (2015) Reactions to the Future: The Chronopolitics of Prevention and Preemption. Nanoethics 9, 165–177. doi: 10.1007/s11569-015-0231-4

Kane R (1996) The significance of free will. Oxford University Press, New York

Kane R (2007) Libertarianism. In: Fischer JM, Kane R, Pereboom D, Vargas M (eds) Four views on free will. Blackwell Publishing, Malden, Oxford, pp 5–43

Kant I (1996) Grundlegung zur Metaphysik der Sitten - Kritik der praktischen Vernunft. Werkausgabe Bd. VII, 13th edn. Suhrkamp Taschenbuch Wissenschaft. Suhrkamp, Frankfurt am Main

Kant I (2009) Kritik der reinen Vernunft: Band 1. Werkausgabe Band III, 1st edn. Suhrkamp Taschenbuch Wissenschaft. Suhrkamp, Frankfurt am Main

Keil G (2000) Handeln und Verursachen. Philosophische Abhandlungen, Bd. 79. Klostermann, Frankfurt am Main

Keil G (2007) Willensfreiheit. Grundthemen Philosophie. De Gruyter, Berlin, New York

Keil G (2009) Willensfreiheit und Determinismus. Grundwissen Philosophie. Reclam, Stuttgart

Keil G (2011) Keine Strafe ohne Schuld, keine Schuld ohne freien Willen? In: Schnädelbach, Herbert, Hastedt, Heiner, Keil G (eds) Was können wir wissen, was sollen wir tun?: Zwölf philosophische Antworten, 2nd edn. Rowohlt, Hamburg, pp 147–167

Kim J (2006) Philosophy of mind, 2nd edn. Westview Press, Boulder, Colorado

Kurzweil R (1999) The age of spiritual machines: When computers exceed human intelligence. Penguin, New York

Laplace PSd (1932) Philosophischer Versuch über die Wahrscheinlichkeit. Ostwald's Klassiker der exakten Wissenschaften, vol 233. Akademische Verlagsgesellschaft, Leipzig

Lehrer K (1964) ‚Could' and Determinism. Analysis 24, 159–160. doi: 10.1093/analys/24.4.159

Lenk H (1993a) Ethikkodizes für Ingenieure: Beispiele der US-Ingenieurvereinigungen. In: Lenk H, Ropohl G (eds) Technik und Ethik, 2nd edn. Reclam, Stuttgart, pp 194–221

Lenk H (1993b) Über Verantwortungsbegriffe und das Verantwortungsproblem in der Technik. In: Lenk H, Ropohl G (eds) Technik und Ethik, 2nd edn. Reclam, Stuttgart, pp 112–148

Lenk H (1994) Zwischen Technokatastrophen und Hoffnungen. In: Lenk H (ed) Macht und Machbarkeit der Technik. Reclam, Stuttgart, pp 35–45

Lenk H (2007) Global technoscience and responsibility: Schemes applied to human values, technology, creativity and globalisaton. Philosophy in International Context, Bd. 3. Lit Verlag, Berlin, London

Lenk H (2009) Zur Verantwortung des Ingenieurs. In: Maring M (ed) Verantwortung in Technik und Ökonomie. KIT Scientific Publishing, Karlsruhe, pp 9–36

Lenk H, Maring M (2001) Verantwortung. In: Ritter J, Gründer K, Gabriel G (eds) Historisches Wörterbuch der Philosophie: Band 11 (U-V). Schwabe, Basel, pp 203–221

Lenk H, Ropohl G (eds) (1993) Technik und Ethik, 2nd edn. Reclams Universal-Bibliothek. Reclam, Stuttgart

Locke J (2008) An essay concerning human understanding. Oxford University Press, Oxford

Lösch A (2006) Means of Communicating Innovations: A Case Study for the Analysis and Assessment of Nanotechnology's Futuristic Visions. Science, Technology & Innovation Studies 2, 103–125

Louden RB (2007) On Some Vices of Virtue Ethics. In: Crisp R, Slote MA (eds) Virtue ethics. Oxford University Press, Oxford, pp 201–216

Mackie JL (1990) Ethics: Inventing right and wrong. Penguin, London

Maring M (2001) Kollektive und korporative Verantwortung: Begriffs- und Fallstudien aus Wirtschaft, Technik und Alltag. Forum Humanität und Ethik, Bd. 2. Lit Verlag, Münster

Martin MW (2002) Personal Meaning and Ethics in Engineering. Science and Engineering Ethics 8, 545–560

Martin MW (2006) Moral Creativity in Science and Engineering. Science and Engineering Ethics 12, 421–433

Martin MW (2007) Creativity: Ethics and excellence in science. Lexington Books, Lanham, MD

Martins EC, Terblanche F (2003) Building organisational culture that stimulates creativity and innovation. European Journal of Innovation Management 6, 64–74. doi: 10.1108/14601060310456337

May L (1991) Metaphysical guilt and moral taint. In: May L, Hoffman S (eds) Collective responsibility: Five decades of debate in theoretical and applied ethics. Rowman & Littlefield, Lanham, Maryland, pp 239–254

Mayr E (1979) Evolution und die Vielfalt des Lebens. Springer, Berlin, Heidelberg

McCray P (2012) California Dreamin': Visioneering the Technological Future. In: Janssen V (ed) Where minds and matters meet: Technology in California and the West. University of California Press, Berkeley, pp 347–378

McCray P (2013) The visioneers: How a group of elite scientists pursued space colonies, nanotechnologies, and a limitless future. Princeton University Press, Princeton

McGinn C (2000) Ethics, evil, and fiction. Clarendon Press, Oxford

Mele AR (1999) Ultimate Responsibility and Dumb Luck. In: Paul EF, Miller FD, Paul J (eds) Responsibility. Cambridge University Press, Cambridge, New York, pp 274–293

Meyer SS (1999) Fate, Fatalism, and Agency in Stoicism. In: Paul EF, Miller FD, Paul J (eds) Responsibility. Cambridge University Press, Cambridge, New York, pp 250–273

Michael M (2000) Futures of the Present: From Performativity to Prehension. In: Brown N, Rappert B, Webster A (eds) Contested futures: A sociology of prospective techno-science. Ashgate, Aldershot, Burlington, pp 21–39

Miller CA, Bennett I (2008) Thinking longer term about technology: Is there value in science fiction-inspired approaches to constructing futures? Science and Public Policy 35, 597–606. doi: 10.3152/030234208X370666

Mitcham C (1987) Responsibility and Technology: The Expanding Relationship. In: Durbin PT (ed) Technology and responsibility. D. Reidel Publishing Company, Dordrecht, Boston, Lancaster, Tokyo, pp 3–39

Moore GE (1978) Freier Wille. In: Pothast U (ed) Seminar, freies Handeln und Determinismus, 1st edn. Suhrkamp, Frankfurt am Main, pp 142–156

Nagel T (1986) The view from nowhere. Oxford University Press, New York

Nagel T (1991a) Equality and partiality. Oxford University Press, Oxford

Nagel T (1991b) Moral Luck. In: Mortal questions. Cambridge University Press, London, pp 24–38

Nagel T (1991c) Subjective and Objective. In: Mortal questions. Cambridge University Press, London, pp 196–213

Nagel T (1991d) The fragmentation of value. In: Mortal questions. Cambridge University Press, London, pp 128–141

Nichols S (2006) Folk Intuitions on Free Will. Journal of Cognition and Culture 6, 57–86. doi: 10.1163/156853706776931385

Nida-Rümelin J (2011) Verantwortung. Reclams Universal-Bibliothek. Reclam, Stuttgart

Nordmann A (ed) (2006) Nanotechnologien im Kontext: Philosophische, ethische und gesellschaftliche Perspektiven. Akademische Verlagsgesellschaft, Berlin

Nordmann A (2011) The Age of Technoscience. In: Nordmann A, Radder H, Schiemann G (eds) Science transformed?: Debating claims of an epochal break. University of Pittsburgh Press, Pittsburgh, pp 19–30

Nordmann A (2013a) (Im)plausibility². International Journal of Foresight and Innovation Policy 9, 125–132. doi: 10.1504/IJFIP.2013.058612

Nordmann A (2013b) Visioneering Assessment: On the Construction of Tunnel Visions for Technovisionary Research and Policy. Science, Technology & Innovation Studies 9, 89–94

Nowotny H (1995) The dynamics of innovation: On the multiplicity of the new. Public lectures / Collegium Budapest, Institute for Advanced Studies, vol 12. Collegium Budapest, Budapest

Nowotny H (2006) Introduction: The Quest for Innovation and Cultures of Technology. In: Nowotny H (ed) Cultures of technology and the quest for innovation. Berghahn Books, New York, pp 1–26

Nozick R (2006) The examined life: Philosophical meditations, 1st edn. Simon & Schuster, New York

Nussbaum MC (1992) Introduction: Form and Content, Philosophy and Literature. In: Nussbaum MC (ed) Love's knowledge: Essays on philosophy and literature. Oxford University Press, New York, pp 3–53

Nye DE (2006) Technology matters: Questions to live with. MIT Press, Cambridge, Mass

Overall C (2005) Aging, death, and human longevity: A philosophical inquiry. University of California Press, Berkeley, London

Paschen H, Coenen C, Fleischer T (2004) Nanotechnologie: Forschung, Entwicklung, Anwendung. Springer, Berlin, Heidelberg, New York

Pauen M (2004) Illusion Freiheit?: Mögliche und unmögliche Konsequenzen der Hirnforschung. Fischer, Frankfurt am Main

Pereboom D (2007) Hard Incompatibilism. In: Fischer JM, Kane R, Pereboom D, Vargas M (eds) Four views on free will. Blackwell Publishing, Malden, Oxford, pp 85–125

Pettit P (2007) Responsibility Incorporated. Ethics 117, 171–201

Pritchard MS (2001) Responsible engineering: The importance of character and imagination. Science and Engineering Ethics 7, 391–402. doi: 10.1007/s11948-001-0061-3

Putnam H (1981) Meaning and the moral sciences. International Library of Philosophy and Scientific Method. Routledge & Kegan Paul, Boston, London, Henley

Rachels J (1993) The elements of moral philosophy, 2nd edn. The Heritage Series in Philosophy. McGraw-Hill, New York

Railton P (1984) Alienation, Consequentialism, and the Demands of Morality. Philosophy & Public Affairs 13, 134–171

Rau J (1978) Ein tödliches Kalkül: Der Ford-Konzern unter Beschuß. Die Zeit

Rescher N (1960) Choice without preference: A Study of the History and of the Logic of the Problem of "Buridan's Ass". Kant-Studien 51, 142–175. doi: 10.1515/kant.1960.51.1-4.142

Rip A (2006) Folk Theories of Nanotechnologists. Science as Culture 15, 349–365. doi: 10.1080/09505430601022676

Rip A (2012) The Context of Innovation Journeys. Creativity and Innovation Management 21, 158–170. doi: 10.1111/j.1467-8691.2012.00640.x

Rip A, Voß J-P (2013) Umbrella Terms as Mediators in the Governance of emerging Science and Technology. Science, Technology & Innovation Studies 9, 39–59

Rogers EM (1995) Diffusion of innovations, 4th edn. Free Press, New York

Ropohl G (1991) Technologische Aufklärung: Beiträge zur Technikphilosophie, 1st edn. Suhrkamp Taschenbuch Wissenschaft. Suhrkamp, Frankfurt am Main

Ropohl G (2009) Verantwortung in der Ingenieurarbeit. In: Maring M (ed) Verantwortung in Technik und Ökonomie. KIT Scientific Publishing, Karlsruhe, pp 37–54

Russell P (1990) Hume on Responsibility and Punishment. Canadian Journal of Philosophy 20, 539–564

Russell B (1999) The problems of philosophy. Dover Publications, Mineola, N.Y

Russell P (2008) Hume on Free Will. In: Zalta EN (ed) The Stanford Encyclopedia of Philosophy, Fall 2008

Sand M (2016) Technikvisionen als Gegenstand einer Ethik von Innovationsprozessen. In: Maring M (ed) Zur Zukunft der Bereichsethiken: Herausforderungen durch die Ökonomisierung der Welt. KIT Scientific Publishing, Karlsruhe, pp 333–354

Sand M, Jongsma KR (2016a) Towards an Ageless Society: Assessing a Transhumanist Programme. In: Domínguez-Rué E, Nierling L (eds) Ageing and Technology: Perspectives from the Social Sciences, 1st edn. Transcript, Bielefeld, pp 275–294

Sand M, Jongsma KR (2016b) Why Neural Determinism is not real Determinism and why Mental States cannot act. American Journal of Bioethics (Neuroscience) 7, 205–207. doi: 10.1080/21507740.2016.1244128

Sand M, Jongsma KR (2017) The usual suspects: Why techno-fixing dementia is flawed. Med Health Care and Philos 20, 119–130. doi: 10.1007/s11019-016-9747-9

Sand M, Schneider C (2017) Visioneering Socio-Technical Innovations: A Missing Piece of the Puzzle. Nanoethics 11, 19–29. doi: 10.1007/s11569-017-0293-6

Schick A (2016) Whereto speculative bioethics? Technological visions and future simulations in a science fictional culture. Med Humanit 42, 225–231. doi: 10.1136/medhum_2016-010951

Schienstock G (2009) Path Dependency and Path Creation: Some Theoretical Reflections. In: Bechmann G, Gorokhov V, Stehr N (eds) The social integration of science: Institutional and epistemological aspects of the transformation of knowledge in modern society. Edition Sigma, Berlin, pp 85–99

Schlick M (1962a) Problems of Ethics. Dover Publications, New York

Schlick M (1962b) What is the meaning of "moral"? In: Rynin D (ed) Problems of Ethics. Dover Publications, New York, pp 79–99

Schlick M (1962c) When is a man responsible? In: Rynin D (ed) Problems of Ethics. Dover Publications, New York, pp 143–158

Schmidt T (2013) How to understand the problem of moral luck. In: Kahmen B, Stepanians M (eds) Critical essays on "Causation and responsibility". De Gruyter, Berlin, pp 299–310

Schönecker D, Wood AW (2002) Immanuel Kant „Grundlegung zur Metaphysik der Sitten": Ein einführender Kommentar. UTB, vol 2276. Schöningh, Paderborn, München

Schulte P (2010) Plädoyer für einen physikalistischen Naturalismus. Zeitschrift für philosophische Forschung 64, 165–189

Searle JR (1991) Minds, brains and science: The 1984 Reith Lectures. Penguin Books. Penguin, London

Seebaß G (1993a) Freiheit und Determinismus: Teil 2. Zeitschrift für philosophische Forschung 47, 223–245

Seebaß G (1993b) Freiheit und Determinismus: Teil 1. Zeitschrift für philosophische Forschung 47, 1–22

Seebaß G (1994) Die konditionale Analyse des praktischen Könnens. Grazer philosophische Studien: Internationale Zeitschrift für Analytische Philosophie 48, 201–228

Seebaß G (2014) Das Rätsel der Freiheit: Der Freiheitsbegriff in der Praktischen Philosophie. In: Laube M (ed) Freiheit. Mohr Siebeck, Tübingen, pp 211–232

Selinger E, Ihde D, van de Poel I, Peterson M, Verbeek P-P (2012) Erratum to: Book Symposium on Peter Paul Verbeek's Moralizing Technology: Understanding and Designing the Morality of Things. Chicago: University of Chicago Press, 2011. Philos. Technol. 25, 605–631. doi: 10.1007/s13347-011-0058-z

Shelley MW (1989) Frankenstein: Or, The modern Prometheus. Puffin Classics. Puffin, London

Simakova E, Coenen C (2013) Visions, Hype, and Expectations: a Place for Responsibility. In: Owen R, Bessant JR, Heintz M (eds) Responsible Innovation:

Managing the responsible emergence of science and innovation in society. John Wiley & Sons, Chichester, UK, pp 241–266

Simons K (1999) Negligence. In: Paul EF, Miller FD, Paul J (eds) Responsibility. Cambridge University Press, Cambridge, New York, pp 52–93

Singer P (1994) Rethinking life and death: the collapse of our traditional ethics. St. Martin's Press, New York

Singer P (2008) Practical ethics, 2nd edn. Cambridge University Press, Cambridge

Slote MA (1992) From morality to virtue. Oxford University Press, New York

Slote MA (2010) Morality not a system of imperatives. In: Selected essays. Oxford University Press, Oxford, pp 138–149

Smilansky S (2000) Free will and illusion. Clarendon Press, Oxford, New York

Smith AM (2005) Responsibility for Attitudes: Activity and Passivity in Mental Life. Ethics 115, 236–271

Smith AM (2007) On Being Responsible and Holding Responsible. The Journal of Ethics 11, 465–484. doi: 10.1007/s10892-005-7989-5

Smith AM (2012) Attributability, Answerability, and Accountability: In Defense of a Unified Account. Ethics 122, 575–589. doi: 10.1086/664752

Specht R (1972) Innovation und Folgelast: Beispiele aus der neueren Philosophie- und Wissenschaftsgeschichte. Problemata, vol 12. Frommann-Holzboog, Stuttgart-Bad Cannstatt

Sternberg RJ (2003) Wisdom, intelligence, and creativity synthesized. Cambridge University Press, Cambridge

Stilgoe J, Owen R, Macnaghten P (2013) Developing a framework for responsible innovation. Research Policy 42, 1568–1580. doi: 10.1016/j.respol.2013.05.008

Stoecker R (1997) Handlung und Verantwortung: Mackie's Rule Put Straight. In: Meggle G, Mundt A (eds) Analyōmen 2: Proceedings of the 2nd conference "Perspectives in analytical philosophy". De Gruyter, Berlin, New York, pp 357–364

Strawson G (2013) The Impossibility of Moral Responsibility. In: Shafer-Landau R (ed) Ethical theory: An anthology, 2nd edn. Wiley-Blackwell, Chichester, West Sussex, Malden, pp 312–316

Strawson PF (1962) Freedom and resentment. Proceedings of the British Academy 48, 1–25

Strawson PF (1985) Skepticism and naturalism: Some varieties. Woodbridge Lectures, vol 12. Columbia University Press, New York

Sturken M, Thomas D (2004) Introduction: Technological Visions and the Rhetoric of the New. In: Sturken M, Thomas D, Ball-Rokeach SJ (eds) Technological visions: The Hopes and Fears That Shape New Technologies. Temple University Press, Philadelphia, London, pp 1–18

Sturken M, Thomas D, Ball-Rokeach SJ (eds) (2004) Technological visions: The Hopes and Fears That Shape New Technologies. Temple University Press, Philadelphia, London

Suppes P (1993) The Transcendental Character of Determinism. Midwest Stud Philos 18, 242–257. doi: 10.1111/j.1475-4975.1993.tb00266.x

Suppes P (1994) Voluntary Motion, Biological Computation, and Free Will. Midwest Stud Philos 19, 452–467. doi: 10.1111/j.1475-4975.1994.tb00298.x

Swierstra T, Jelsma J (2006) Responsibility without Moralism in Technoscientific Design Practice. Science, Technology & Human Values 31, 309–332. doi: 10.1177/0162243905285844

Taylor R (1987) Time and Life's Meaning. Review of Metaphysics 40, 675–687

Taylor R (2013) Determinism and the theory of agency. In: Shafer-Landau R (ed) Ethical theory: An anthology, 2nd edn. Wiley-Blackwell, Chichester, West Sussex, Malden, pp 308–311

Taylor AL (1982) Striking It Rich: A new breed of risk takers is betting on the high-technology future. Time 119(7). http://content.time.com/time/subscriber/article/0,33009,925279,00.html. Accessed 28 March 2018

Thaler RH, Sunstein CR (2009) Nudge: Improving decisions about health, wealth and happiness. Penguin, London

Toepfer G (2013) Evolution, 1. Aufl. Grundwissen Philosophie. Reclam, Stuttgart

Toulmin SE (1961) Foresight and understanding: An enquiry into the aims of science. Indiana University Press, Bloomington

Tugendhat E (2007) Willensfreiheit und Determinismus. In: Liessmann KP (ed) Die Freiheit des Denkens. Zsolnay, Wien, pp 45–67

Tutton R (2016) Wicked Futures: Meaning, Matter, and the Sociology of the Future. The Sociological Review, 1–16. doi: 10.1111/1467-954X.12443

Ulrich P, Thielemann U (1993) How do managers think about market economies and morality?: Empirical enquiries into business-ethical thinking patterns. J Bus Ethics 12, 879–898. doi: 10.1007/BF00871669

Urmson JO (1999) Aristotle's ethics. Blackwell Publishing, Oxford

Urry J (2016) What is the Future?, 1st edn. Polity Press, Cambridge, United Kingdom

Vallor S (2016) Technology and the virtues: A philosophical guide to a future worth wanting. Oxford University Press, New York

van de Poel I (2011) The relation between forward-looking and backward-looking responsibility. In: Vincent NA, van de Poel I, van den Hoven J (eds) Moral Responsibility: Beyond Free Will and Determinism. Springer, Dordrecht, Heidelberg, London, pp 37–52

van de Poel I (2015a) Moral Responsibility. In: van de Poel I, Royakkers L, Zwart SD (eds) Moral Responsibility and the Problem of Many Hands. Taylor & Francis, New York, pp 12–49

van de Poel I (2015b) The problem of the many hands. In: van de Poel I, Royakkers L, Zwart SD (eds) Moral Responsibility and the Problem of Many Hands. Taylor & Francis, New York, pp 50–92

van de Ven, Andrew, Poole MS (eds) (2004) Handbook of organizational change and innovation. Oxford University Press, New York

van Inwagen P (1975) The incompatibility of free will and determinism. Philos Stud 27, 185–199. doi: 10.1007/BF01624156

Vargas M (2007) Revisionism. In: Fischer JM, Kane R, Pereboom D, Vargas M (eds) Four views on free will. Blackwell Publishing, Malden, Oxford, pp 127–165

Various (2009) One Bad Bump… http://content.time.com/time/specials/packages/article/0,28804,1908719_1908717_1908696,00.html. Accessed 28 March 2018

Various (2016) Cheating Death: Science is getting to grips with ways to slow ageing. Rejoice, as long as the side-effects can be managed. The Economist, 7–16. https://www.economist.com/news/leaders/21704791-science-getting-grips-ways-slow-ageing-rejoice-long-side-effects-can-be. Accessed 28 March 2018

Verbeek P-P (2005) What things do: Philosophical reflections on technology, agency, and design, 2nd edn. Pennsylvania University Press, Pennsylvania

Verbeek P-P (2011) Moralizing technology: Understanding and designing the morality of things. University of Chicago Press, Chicago, London

von Schomberg R (2013) A vision of responsible innovation. In: Owen R, Bessant JR, Heintz M (eds) Responsible Innovation: Managing the responsible emergence of science and innovation in society. John Wiley & Sons, Chichester, UK, pp 51–74

von Schomberg R (2015) Responsible Innovation: The New Paradigm for Science, Technology and Innovation Policy. In: Bogner A, Decker M, Sotoudeh M (eds) Responsible Innovation: Neue Impulse für die Technikfolgenabschätzung?, 1st edn. Edition Sigma, Baden-Baden, pp 47–70

Wallace JR (1994) Responsibility and the moral sentiments. Harvard University Press, Cambridge, Mass.

Waller BN (2011) Against moral responsibility. MIT Press, Cambridge, Mass.

Watson G (2004a) Introduction. In: Agency and answerability: Selected essays. Oxford University Press, New York, pp 1–10

Watson G (2004b) Reasons and Responsibility. In: Agency and answerability: Selected essays. Oxford University Press, New York, pp 289–317

Watson G (2004c) Responsibility and the Limits of Evil: Variations on a Strawsonian Theme. In: Agency and answerability: Selected essays. Oxford University Press, New York, pp 219–259

Watson G (2004d) Two Faces of Responsibility. In: Agency and answerability: Selected essays. Oxford University Press, New York, pp 260–288

Weber M (1988) Die »Objektivität« sozialwissenschaftlicher und sozialpolitischer Erkenntnis. In: Weber M, Winckelmann J (eds) Gesammelte Aufsätze zur Sozial- und Wirtschaftsgeschichte. Mohr Siebeck, Tübingen, pp 146–214

Werhane PH, Freeman ER (2010) Corporate Responsibility. In: LaFollette H (ed) The Oxford handbook of practical ethics. Oxford University Press, Oxford, pp 514–538

Wessels U (2011) Das Gute: Wohlfahrt, hedonisches Glück und die Erfüllung von Wünschen. Klostermann RoteReihe, vol 41. Vittorio Klostermann, Frankfurt am Main

Wiggins D (1976) II—Deliberation and Practical Reason. Proc Aristot Soc 76, 29–52. doi: 10.1093/aristotelian/76.1.29

Wiggins D (2006) Ethics: Twelve lectures on the philosophy of morality. Penguin, London

Williams B (1981a) Moral luck: Philosophical papers 1973-1980. Cambridge University Press, Cambridge

Williams B (1981b) Moral luck. In: Moral luck: Philosophical papers 1973-1980. Cambridge University Press, Cambridge, pp 20–39

Williams B (1981c) Persons, character and morality. In: Moral luck: Philosophical papers 1973-1980. Cambridge University Press, Cambridge, pp 1–19

Williams B (1997) Morality: An introduction to ethics, 3rd edn. Canto. Cambridge University Press, Cambridge

Williams B (2006) Ethics and the limits of philosophy. Routledge, London

Wolf S (1990) Freedom within reason. Oxford University Press, New York

Wolf S (1997) Happiness and Meaning: Two Aspects of the Good Life. Soc Phil Pol 14, 207–225. doi: 10.1017/S0265052500001734

Wolf S (2010) Meaning in life and why it matters. The University Center for Human Values Series. Princeton University Press, Princeton

Wolf S (2013) Sanity and the Metaphysics of Responsibility. In: Shafer-Landau R (ed) Ethical theory: An anthology, 2nd edn. Wiley-Blackwell, Chichester, West Sussex, Malden, pp 330–339

Wollheim R (1999) The thread of life, 1st edn. The William James Lectures. Yale University Press, New Haven

Wright GH von (1974) Causality and determinism. Woodbridge Lectures. Columbia University Press, New York

Ziman J (2000) Evolutionary Models for Technological Change. In: Ziman J (ed) Technological innovation as an evolutionary process. Cambridge University Press, Cambridge, pp 3–12

Zingerle A (1976) Innovation. In: Ritter J, Gründer K, Gabriel G (eds) Historisches Wörterbuch der Philosophie: Band 4 (I-K). Schwabe, Basel, pp 511–519